VENEZUELA

POLITICS IN LATIN AMERICA
A HOOVER INSTITUTION SERIES

General Editor, **Robert Wesson**

Copublished with Hoover Institution Press,
Stanford University, Stanford, California

VENEZUELA

Politics in a Petroleum Republic

David Eugene Blank

PRAEGER

PRAEGER SPECIAL STUDIES • PRAEGER SCIENTIFIC

New York • Philadelphia • Eastbourne, UK
Toronto • Hong Kong • Tokyo • Sydney

Library of Congress Cataloging in Publication Data

Blank, David Eugene.
Venezuela

Bibliography: p.
Includes index.
1. Venezuela—Politics and government—1958–
2. Petroleum industry and trade—Venezuela.
3. Petroleum industry and trade—Government policy—Venezuela.
I. Title. II. Series.
JL3881.B57 1984 987'.0633 83-24469
ISBN 0-03-069792-1 (alk. paper)

Published and Distributed by the Praeger Publishers Division (ISBN Prefix 0-275) of Greenwood Press, Inc., Westport, Connecticut

Published in 1984 by Praeger Publishers
CBS Educational and Professional Publishing,
a Division of CBS Inc.
521 Fifth Avenue, New York, NY 10175 USA

56789 052 987654532

Printed in the United States of America
on acid-free paper

To Jennifer Ethel Blank

PREFACE

This work represents the summing up of a generation of professional interest in the politics of Venezuela and my second attempt to produce a systematic explanation of that nation to North American readers. These readers are invited to compare the evolution of my own perspective over the decade that has separated this volume from my 1973 *Politics in Venezuela*.

I would like to express my gratitude to the Foreign Area Fellowship Program, The American Philosophical Society, and the Research Committee of the College of Arts and Sciences of the University of Louisville for supporting my many field research trips to Venezuela between 1966 and 1981. Without the financial support of these agencies this volume would not have been possible.

My guarantee of anonymity to the many Venezuelan informants who took time out of their busy schedules to answer my questions prevents me from listing their names in most cases. However, a number of Venezuelans have been kind enough over the years to critique my writings on their nation, and I would like to acknowledge my debt to them. The following list represents individuals with diverse party and institutional affiliations whose comments have been especially helpful: Valmore Acevedo Amaya, Luis E. Berrizbeitia, Marco Tulio Bruni Celli, José Antonio Gil-Yepes, R. A. Irving Jahn, and José A. Silva Michelena.

A number of U.S. colleagues have also made incalculable contributions to this book through their comments on tentative versions of several chapters. A special debt is owed to Robert S. Alexander, Kenneth Coleman, Edward Thompson, David J. Myers, Gene Bigler, John D. Martz, Lowell Newton, and Franklin Tugwell. A special expression of gratitude is also due to Robert Wesson and to the Hoover Institution for giving me complete scholarly freedom to make my own mistakes in writing this book free from any institutional pressures.

Sue Pearcy and Sharon Mills of the University of Louisville did a fine job of typing the manuscript. Ed Herzer and Michelle Ising of the Graphics and Photography Department of the University of Louisville prepared all the charts and the map under a most pressing time schedule. Finally, Karen Hansen and Laura Jasper of Praeger Special Studies did an excellent job of editing.

EDITOR'S FOREWORD

Venezuela is one of the Latin American countries of most importance to the United States, and we are especially glad to present a systematic treatment of its politics. David Blank knows Venezuela not only from many years of study but also from extended investigations on the scene, and he is well-qualified to interpret its institutions.

Aside from its intrinsic importance, Venezuela interests us for both material and intellectual reasons. It is, along with Mexico, a reliable hemispheric source of petroleum, and it is the best functioning democracy of Latin America after diminutive Costa Rica. Professor Blank concentrates on these two aspects, analyzing the problems of control of the petroleum industry and the reality and limitations of Venezuelan democracy, weaving together the themes of management and populism. The picture that emerges encourages us to believe that democracy is indeed possible in the Latin American environment, albeit within restrictions imposed by the political reality.

Robert Wesson

CONTENTS

VENEZUELA

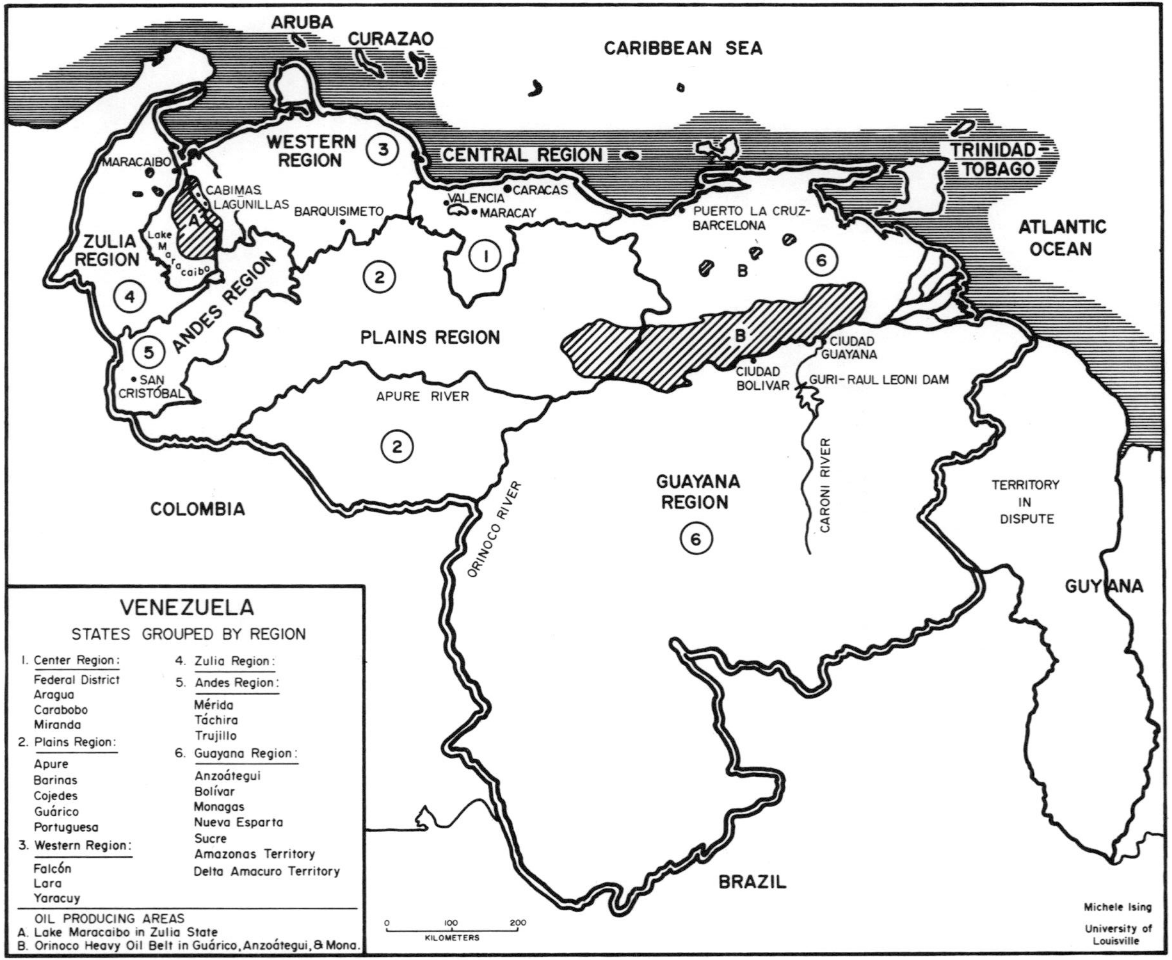
ARUBA
CURAZAO
CARIBBEAN SEA
WESTERN REGION
3
CENTRAL REGION
TRINIDAD-TOBAGO
ATLANTIC OCEAN
MARACAIBO
CABIMAS
LAGUNILLAS
BARQUISIMETO
VALENCIA
MARACAY
CARACAS
PUERTO LA CRUZ-BARCELONA
ZULIA REGION
Lake Maracaibo
A
4
ANDES REGION
5
SAN CRISTOBAL
1
2
PLAINS REGION
B
6
CIUDAD GUAYANA
CIUDAD BOLIVAR
GURI-RAUL LEONI DAM
APURE RIVER
2
COLOMBIA
ORINOCO RIVER
GUAYANA REGION
6
CARONI RIVER
TERRITORY IN DISPUTE
GUYANA
BRAZIL
VENEZUELA
STATES GROUPED BY REGION
1. Center Region:
Federal District
Aragua
Carabobo
Miranda
2. Plains Region:
Apure
Barinas
Cojedes
Guárico
Portuguesa
3. Western Region:
Falcón
Lara
Yaracuy
4. Zulia Region:
5. Andes Region:
Mérida
Táchira
Trujillo
6. Guayana Region:
Anzoátegui
Bolívar
Monagas
Nueva Esparta
Sucre
Amazonas Territory
Delta Amacuro Territory
OIL PRODUCING AREAS
A. Lake Maracaibo in Zulia State
B. Orinoco Heavy Oil Belt in Guárico, Anzoátegui, & Mona.
0
100
200
KILOMETERS
Michele Ising
University of Louisville

1

INTRODUCING VENEZUELA'S POLITICAL SYSTEM

The heightened visibility of Venezuela largely reflects the fact that the nation has been and should continue to be a secure and friendly Western Hemispheric producer of oil over the next generation or two. The critical importance of Venezuelan oil is underlined by the continuing probability of production disruptions in the Persian Gulf region as well as the prospect of a major political explosion in the Middle East.

Venezuela, however, should not be treated exclusively in terms of its oil. Interest in Venezuela also reflects the fact that since 1958 that nation has experimented with one of the Third World's few constitutional democracies. The future success of democracy in meeting the challenge of development is crucial not only to Venezuela but also to the United States, because it could provide a model of a viable democratic alternative for other Third-World nations. Venezuela has already assumed the role of being an honest broker in mediating the highly emotional issues involved in the North-South dialogue between the industrial democracies of the First World and its fellow Third-World states. This mediator role has become especially important in the critical Caribbean basin.

This book has been designed as a guide to the dynamics and meaning of Venezuela's democratic experiment. The presentation of the clash of ideas and theoretical frameworks as they relate to Venezuela comprises an integral part of the book because objective facts often do not speak for themselves; they require incorporation into some kind of conceptual framework and normative perspective in order to have meaning. This work builds upon the considerable contribution a host of scholars has already made to our understanding of politics in Venezuela.[1] The debt owed these friends and colleagues is not diminished by the fact that my own scholarship has at times led me to conclusions different from theirs.

Contemporary Venezuela is a synthesis of more than 150 years of historical evolution as an independent nation with four centuries of hispanic cultural formation, on one hand, and the explosive intrusion of oil, commercially produced since 1917, on the other. This synthesis of internal, or endogenous, development and external, or exogenous, causation has meant that over the past 60 years philosophical notions, production techniques, and political organizational forms that were once considered foreign have become incorporated into the evolving Venezuelan way of life. This loss of economic self-sufficiency and autonomous, political, inner direction reflects the increased global interdependence affecting all. It is not unique to Venezuela.

Petroleum has been the pivot around which Venezuelan politics have revolved for most of the twentieth century. The politics of oil in Venezuela have had domestic and international dimensions as that nation moved from being a passive supplicant of foreign-oil-company investment during a period of traditional warlord, or *caudillo*, rule to the present time in which an increasingly politically conscious nation manages its nationalized oil industry with the intent of generating a stable and integrated development.

Venezuelan oil has also provided the basis for the nation's newly assertive international presence. Describing and analyzing the continued evolution of the democratic management of oil is one of the twin focuses of this book. The second is the politics of "sowing oil" profits to bring out economic modernization, social justice, and political stability. Venezuela has sought to create an integrated and balanced economy that will be able to prosper in a prospective post-petroleum era ever since Alberto Adriani and Uslar Pietri, advisers to President López Contreras in the late 1930s, first articulated the rallying cry, "*sembrar el petroleo*" (sow the oil).[2]

Since the 1973 Organization of Petroleum Exporting Countries (OPEC) price revolution, which more than quadrupled Venezuela's government income, the nation has failed to buy its way into modernity and stability. Despite official pledges made in the mid-1970s to manage "abundance with the mentality of scarcity," an avalanche of public spending over the past few years has perhaps created more problems than it has solved.[3] Runaway inflation and a soaring crime rate have traumatized and terrorized the urban population. Domestic agriculture has deteriorated as more and more Venezuelans seek the good life in the cities and on the public payroll. In 1980, Venezuela had to once again import more than 50 percent of its basic food staples. Soaring imports and the need to service an exploding foreign debt, which in 1981 reached 21 percent of the annual national budget, threaten to cancel many of the benefits of oil exports. In 1982, the world oil glut cut into Venezuela's principal source of revenue.

Two leaders—the late Juan Pablo Pérez Alfonzo and former President Rafael Caldera—who played important roles in the post-1958 development of democracy in Venezuela articulated Venezuelan ambivalence toward its oil wealth. Writing in the mid-1970s during the petroleum boom, Juan Pablo Pérez Alfonzo warned of the corrupting influence sudden wealth provided. "Venezuela," he wrote "could die of indigestion as certainly as from hunger."[4] Juan Pablo Pérez Alfonzo was a leader of one of the nation's principal political parties, Democratic Action (AD), and of its oil ministry under the late President Rómulo Betancourt (1959–64).

In similar fashion, Rafael Caldera, a founder of Venezuela's other principal political party, the Social Christian Party (COPEI), president 1969–74, and candidate for the presidency in 1983, summarized his view of the negative impact of the oil industry by pointing out that it distorted the territorial organization and social structure of the nation, disrupted its traditional agriculture and industry, and destroyed part of its environment. While the government has become largely dependent on oil revenue for 70 percent of its taxes, that industry only provides 0.5 percent of national employment. In 1981, former President Caldera wrote, "It will no longer be sufficient [for Venezuela] to merely sow petroleum, but it will increasingly be necessary [for Venezuela] to dominate it [*dominar el petróleo*].[5]

COPING WITH DEPENDENCY AND THE LEGACY OF THE PAST

It is incumbent on the author to present his own perspective and normative frame of reference before subjecting the reader to his description and analysis. The primary purpose of this research perspective is to provide a model of Venezuelan politics that will be meaningful to the North American reader. The work of scholars has provided two quite competitive models with which to describe Venezuelan reality.[6] One group of scholars has developed the modernization or development perspective, which has focused on the internal structural and value changes Venezuela and other developing nations must go through in order to overcome the historical legacy of traditional authoritarianism, personalism, and paternalism.

A second group of scholars has developed the dependency perspective, which has focused on Venezuela's inclusion into a U.S.-dominated system of global capitalism. The work of the second group focuses on the need to lessen Venezuela's dependence on that system and to create a more autonomous nation. For the second group, internal problems of authoritarianism and paternalism are merely reflective of national dependency.

While the author incorporates a number of key elements of both perspectives in his acknowledgment of the importance of both cultural and external constraints on contemporary Venezuela, he strongly rejects the determinative quality of both, which tends to minimize the relevance of indigenous politics. Both perspectives, in their more extreme versions, tend to dismiss the ability of democratic politics to have much of an impact on either the legacy of the past or on the international linkages of the present. A key assumption of this book is that democratic politics in Venezuela, with all their defects and failings, are a significant and autonomous part of its national reality.

While not wishing in this introductory chapter to impose a theoretical polemic on readers, the author would briefly like to illustrate a potential influence of the debate between the development perspective and the dependency perspective on our perception of Venezuela's political reality. Conservative writers have tended to justify their support of traditional autocrats and modernizing authoritarians because this political style is said to be more attuned to Latin America's culture of personalism, paternalism, and patron-clientelism. Venezuela, despite its democratic forms, shares these cultural characteristics. Should one therefore assume, with Jeane Kirkpatrick, that the clamor and clang of its electoral process are mere "democratic facade-elections, political parties, and a fairly broad participation . . . modified by varying degrees of fraud, intimidation, and restrictions on who may participate. Corruption is endemic. . . . Competition for influence proceeds by whatever means are at hand. . . ."[7]

While conservative writers tend to downgrade the viability of democracy because of its alleged cultural incompatibility and the ease with which that culture rejects it, radical writers tend to dismiss it as a mere illusion because of the awesome influence of the United States on Venezuela. Democratic politics in Venezuela are in part an illusion allegedly because of the strict limits U.S. capitalism places on them. "North American military missions did not confine their advise to warfare techniques. . . . Anti-Communist ideology . . . was introduced into Latin American troop indoctrination," converting national armies into "occupation forces."[8] According to the author of this quotation, J.A. Silva Michelena, a more subtle domination of Venezuela has also resulted from the cultural penetration of U.S. transnationals, which have denationalized national industry, dominated mass media, and stultified indigenous technology.

The suggestion that democracy in Latin America is a mere facade comes from an article written in 1980 by Jeane Kirkpatrick, who has since become U.S. ambassador to the United Nations. The reference to Venezuelan democracy as an illusion was part of a title of a 1971 book written by José Agustin Silva Michelena, now director of academic studies at the Venezuela's Central University Center for the Study of Development and a supplementary national deputy, or congressman, elected in 1978 from the slate of the growing Socialist Party (MAS). Ambassador Kirkpatrick and Professor Silva Michelena had the pleasure of briefly meeting each other at a seminar on U.S.-Venezuelan relations held at the Woodrow Wilson Institute in Washington, D.C., in November 1981. They may have had the opportunity to mention to each other that in their own way each is now working to make Venezuela's democratic reality much less of an illusion or facade, despite the continuation of close ties both to the United States and to its Latin culture.

These citations from the writings of Jeane Kirkpatrick and Silva Michelena have been used to allow the author the opportunity to point out that U.S. social-science writings are replete with an insiduous thesis that authoritarianism (of either a radical revolutionary or conservative bureaucratic variety) is a more natural and suitable political style for developing nations such as Venezuela than is pluralistic democracy. Suggesting that "corn ball" democratic politics, in their own nonspectacular, not all that inspiring way, may overcome the limits of history, culture, and international linkages is the message of this book.

The above polemical detour and the author's own confession that he expects to see Venezuela's nonspectacular democracy survive the challenges of the 1980s was written with the recognition that in early 1983 there are many indicators that the oil boom of the mid-1970s is over. There is the possibility that for the next three to five years Venezuela's populist democracy will have to muddle through with a declining revenue from oil. The 1980–82 drop in the world price of oil has approached 20 percent. This will unquestionably present a grave new challenge to Venezuela's democratic institutions.

VENEZUELA'S TRIPARTITE POLITICAL SYSTEM

In no political system is the democratic process in its multiple forms the only politics. The need of mass-based, partisan-organized, representative democracy to coexist with other styles of politics reflects the reality of just about all nations. Venezuela's contemporary democratic experiment dates back only to 1958. It has had to overcome not only the legacy of its own history but also the challenge of transnational forces that have at various times attempted to destroy or restrict its future.

As late as 1967, Cuba's Fidel Castro praised the effort of a small band of radical insurgents to overthrow the system, even if that goal was to be achieved by means of random violence and assassinations. This Cuban intrusion was coincident with a campaign of the large transnational oil companies to sabotage Venezuela's economic-development policy making by means of their own disinvestment and the encouragement of a massive flight of indigenous private capital. This and other features of Venezuela's political reality have made it critical that representative democracy coexists with other forms of political representation and policy making. The basic outline of Venezuela's tripartite politics outlined in the author's 1973 volume is retained for this book.[9]

There are two complementary ways of illustrating this differentiated political system. The first model is that of a disassociation of power, utilized by Silva Michelena in his 1971 book.[10] Three parallel, political subsystems utilizing different political resources—representative democracy, planning, praetorianism—share the pyramid of

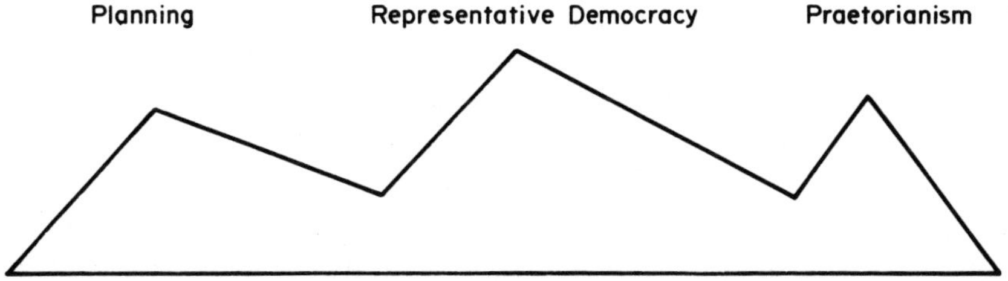

Figure 1.1. Venezuela's Power Pyramid

power in Venezuela. Rather than one, focused apex, the pyramid of authoritative decision-making power in Venezuela has, according to this model, three parallel peaks, as Figure 1.1 illustrates.

Representative democracy is the central political process, and its peak is somewhat higher than that of the other political subsystems. The more the politics of planning and praetorianism dominate representative democracy—lowering the height of its authoritative position and moving it off center—the more it resembles an illusion or facade. The fact that representative government must negotiate with the other power centers does not, however, automatically negate its significance.

The subsystem that incorporates the political participation of functionally important business groups, key state corporations, and the domestic presence of the transnational corporations, all of whose power is independent of the electoral process, has been called the politics of planning. The author prefers this label to the one of "corporatism," currently in vogue, because it emphasizes the ability of planning and partisan politics to coexist, while corporatism implies a basic incompatibility. Direct, functional representation of technical know-how or economic strength is, in the author's view, one of the legitimate forms of participation in a pluralistic democracy, although it may not have formal ties to the political party system. The question of when a planning dialogue or negotiations between an elected government and key interest groups is compatible with or inimical to democracy is a question asked of the industrial democracies of the developed First World as well as of the Third World, such as Venezuela.[11] Planning results from a recognition of the complex, technical reality of public policy making in the twentieth century rather than from any retreat from democracy.

The second nonelectoral political subsystem has been called praetorianism. This concept replaces the more narrowly construed "politics of violence" used by this author in his 1973 volume. Praetorianism refers to the political style that uses the mobilization of force and violence to influence public decision making. In praetorian politics, according to S. P. Huntington, the end justifies the means because there are no generally accepted rules of the game; businessmen bribe, students riot, and revolutionaries of the military feel free to seize power.[12] The broader definition of praetorianism that indicates that all social groups in society tend to resort to force and violence and to violate whatever may exist of a legal system in order to achieve their goals is used in this volume rather than the more narrow one that refers exclusively to military intervention.

In maneuvering through its own inherent conflicts and the constraints placed on it by national culture (as expressed by praetorianism) and by external economic linkages (as expressed in the need for planning), democracy in Venezuela has developed a confusing, but apparently necessary, split personality. To some, it presents a populist face, to others, it has a reformist personality.

The literature on populism in Latin America, as elsewhere, describes a most ambiguous concept.[13] Populism has been described in terms of social psychology, of ideology, and of its charismatic leadership style. In contemporary Venezuela, yesterday's anti-status quo, grass-roots movements have become today's establishment political parties: coalitions of disparate interests and classes, often lacking any coherent ideology and focusing on creating a patronage machine and "throwing money at social problems." Because of its ability to coexist with, and perhaps be absorbed by, the old order, populism is increasingly used as a pejorative.

Authoritarian revolutionaries such as Cuba's Fidel Castro have justified using violence to destroy fragile democracies such as that of Venezuela on the grounds that the representative process and institutions are incapable of executing basic social and economic reforms. According to these authoritarians, achieving real social justice and national liberation requires a catastrophic upheaval: To make an omelet, one must break an egg.

The democratic process or political subsystem in Venezuela has indeed become part of the ongoing, evolving social order. Unlike the revolutionary process, democracy is indeed not at war with its society and culture. Nor does it envision any grand transformation of its society and culture.

Although populist muddling-through has indeed not brought about any dramatic "leap to freedom," even some of its most severe critics acknowledge the crucial mediating and brokerage role that populist politics have played in preventing a violent clash between society's corporatist and praetorian tendencies. Despite the often corrupt and clientelist reality of this frequently maligned democratic "persona," it has been responsible for making Venezuela's peasants, organized workers, and poor in general self-confident and loyal participants in the nation's politics. The literature on Venezuela is replete with references to the fact that, thanks to populist democracy, the poor and relatively poor cannot be easily pushed around. Democratic politics have enabled the poor to turn to politicians in times of need.

Foreign scholars might deprecate this critical role; however, many Venezuelans, who know how unique their nation is with its climate of civil liberties and political civility, do not underestimate its importance. While old, established forces have not been eliminated from Venezuela's politics, they have been domesticated. By accepting the democratic rules of the game, these once-dominant forces have accepted much change. Populism, with its slow and stealthful ways, may alienate power groups, but it has not threatened them to the point at which they will turn against democracy.

The governing of a restless nation such as Venezuela over the past generations has indeed driven democracy to distraction. The writer, however, accepts the suggestion of Albert O. Hirschman that the difference between the stealthful ways of democratic populism muddling-through and its wheeling and dealing and the more purposive and direct reform mongering of the democratic reform "persona" may be largely semantic.[14] Democratic reform may talk of more substantial change than it can deliver, while populist wheeling and dealing may achieve more substantial change than it intends.

Political scientists Wynia and Nelson have conceptually distinguished democratic populism from democratic reform.[15] The victory of a belief in the possible rather than striving for the preferred, the triumph of organizational needs over party program, and the triumph of leadership careers over internal party democracy have all distinguished populism from reformism. Whereas democratic reform argues for the slow creation of autonomous and assertive grass-roots mass organizations and electoral constituencies, the reality of governing a developing nation has led politicians "to throw money at the nations' problems," seeking merely to win the next elections by offering public works, patronage, and paternalistic welfare programs. The goal of electoral democracy may be the creation of a self-conscious and responsible party constituency; its reality is often the interdependent exchange between party and electoral clienteles.

The second model of Venezuela's tripartite politics may better illustrate this stealthful change-agent capability of democratic populism than the more static model of a disassociated power pyramid. In Figure 1.2, Venezuela's tripartite political system is schematically represented as three interlocked rings. Venezuela's electoral system, with its slow and stealthful populism muddling-through style, moves the entire system toward achieving greater social justice and national autonomy.

The planning and praetorian politics tend to resist this tension because of their ties to either transnational forces or to cultural traditions. The actual movement of the system occurs through periodic efforts at tension change and relaxation-consolidation cycles. Movement also occurs along two disjointed, or incremental, zig zag cutting edges of change, along the interfaces of democratic politics with that of planning and praetorianism.

Too much tension change might produce a breakdown of the democratic consensus, while too little may produce a depoliticalization of the masses of people and their encapsulation into a privatized life and political cynicism. This is in fact a fairly accurate description of the ongoing democratic dilemma, when to push for change and when to relax.

This book represents the author's long-maturing reaction to Venezuela's politics. There is too much complexity and contradiction in Venezuela to justify the application of simplified explanatory models. While not accepting at face value many of the claims innumerable Venezuelans have asserted regarding their role in the system or that of their party or organization or, for that matter, electoral democracy, this writer wishes to again assert his advocacy of the belief that the worst democracy is preferable to the best dictatorship. Venezuela's democracy, with all its faults, is far from being the worst.

Scholars, like tourists, pick and choose only certain aspects of the nation they study. In concluding this introductory chapter, this author states his agreement with Dan Levine's recent (1981) assertion: "Scholars and observers of many kinds have come to Venezuela with very different goals, concepts, and tools of analysis," they have also, to a large degree, "seen what they came to see."[16]

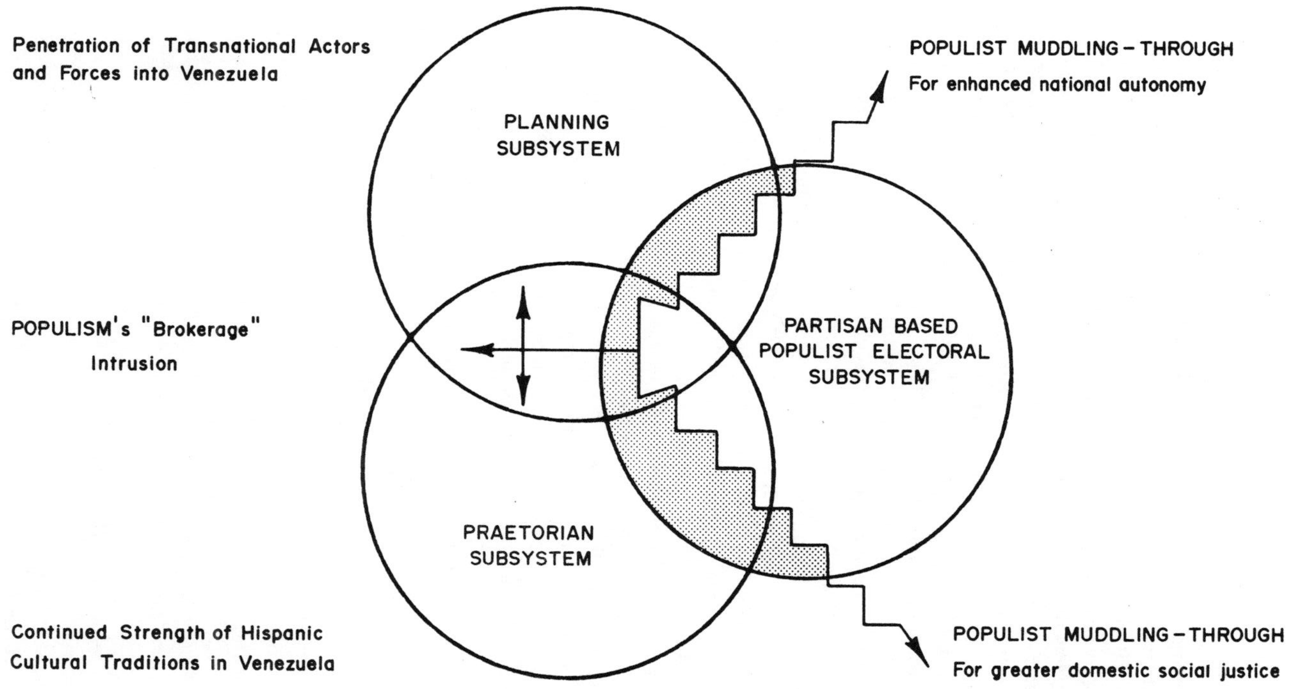

Figure 1.2. Venezuela's Tripartite Political System

NOTES

1. The following works constitute a sample of the literature on Venezuela written over the years from a variety of perspectives. Edwin Lieuwen, *Petroleum in Venezuela: A History* (Berkeley: University of California Press, 1954); Robert J. Alexander, *The Venezuelan Democratic Revolution: A Profile of the Regime of Rómulo Betancourt* (New Brunswick: Rutgers University Press, 1964); John Martz, *Acción Democrática: Evolution of a Modern Political Party in Venezuela* (Princeton: Princeton University Press, 1966); David E. Blank, *Politics in Venezuela* (Boston: Little, Brown, 1973); Daniel H. Levine, *Conflict and Political Change in Venezuela* (Princeton: Princeton University Press, 1973); Franklin Tugwell, *The Politics of Oil in Venezuela* (Stanford: Stanford University Press, 1975); J.A. Silva Michelena, *The Illusion of Democracy in Dependent Nations* (Cambridge: Massachusetts Institute of Technology Press, 1971); John D. Martz and David J. Myers (eds.), *Venezuela: The Democratic Experience* (New York: Praeger, 1977).

2. There is some confusion concerning the origins of the policy of sowing the oil. Tugwell, *The Politics of Oil*, credits Arturo Uslar Pietri, a friend and close associate of Adriani. The author has been assured by José Antonio Giacopini Zarraga, a close friend of and adviser to the president of the Petroleum of Venezuela S.A., that the originator was Adriani.

3. "Unwieldy Wealth: Venezuelan Oil Riches Fail to Trickle Down," *Wall Street Journal*, February 3, 1981, contains the quotation attributed to President Carlos Andrés Pérez (1974–79). The article further states that Pérez "launched big programs in all directions at once," a sort of "pharoanism."

4. Quoted in Jean-Pierre Clerc, "The Difficult Art of Sowing Oil," *Venezuela News* (a documentary service on Venezuela published by the Venezuelan Mission to the United Nations) 2 (May 30, 1977), p. 1.

5. Rafael Caldera, "Prologo," in *Profundizacion de la Nacionalizacion Petrolera Venezolana: Aspectos Economicos, Administrativos y Fiscales*, ed. Jose Ignacio Moreno Leon (Caracas: Edicones Centauro, 1981), p. XIII.

6. J. Samuel Valenzuela and Arturo Valenzuela, "Review Essay: Modernization and Dependency, Alternative Perspectives in the Study of Latin America Underdevelopment," *Comparative Politics* 10 (July 1978).

7. Jeane Kirkpatrick, "U.S. Security & Latin America," *Commentary* 71 (January 1981), p. 34.

8. Silva Michelena, *The Illusion of Democracy*, p. 279.

9. Blank, *Politics in Venezuela*. In 1973, I identified the three Venezuelan political subsystems as partisan-based politics, the politics of planning, and the politics of violence.

10. Silva Michelena, *The Illusion of Democracy*, pp. 57–64.

11. This dilemma in the evolution of contemporary democracy is a central theme of Reginald J. Harrison, *Pluralism and Corporatism: The Political Evolution of Modern Democracies* (Boston: George Allen & Unwin, 1980).

12. Samuel P. Huntington, *Political Order in Changing Societies* (New Haven, Conn.: Yale University Press, 1968), pp. 196–97.

13. Two classic discussions on populism in Latin America are Alistair Hennessy, "Latin America," in *Populism: Its Meaning and National Characteristics*, eds. Ghita

Ionescu and Ernest Gellner (London: Macmillan, 1969), and Torcuato Di Tella, "Populism and Reform in Latin America," in *Obstacles to Change in Latin America*, ed. Claudio Véliz (London: Cambridge University Press, 1965), pp. 47–74. The more recent literature has focused on the distinction between rural and urban populism. See, for example, Ronald C. Newton, "Latin American Populism: Some Notes on Periodization," in *The Continuing Struggle for Democracy in Latin America*, ed. Howard J. Wiarda (Boulder, Colo.: Westview Press, 1980), pp. 71–76.

14. Albert O. Hirschman, *Journeys Toward Progress: Studies of Economic Policy-Making in Latin America* (New York: The Twentieth Century Fund, 1963), p. 251.

15. Gary W. Wynia, *The Politics of Latin American Development* (Cambridge: Cambridge University Press, 1978), and Joan M. Nelson, *Access to Power: Politics and the Urban Poor in Developing Nations* (Princeton: Princeton University Press, 1979).

16. Daniel H. Levine, "Review Essay: Portraits of Venezuela" *Journal of Inter-American Studies and World Affairs* 23 (May 1981): 203–33.

2

THE HISTORICAL AND CULTURAL SETTING

THE LEGACY OF THE PAST

Spanish explorers probing the northern coast of South America during the sixteenth century came across Lake Maracaibo. The Indians living in that area built their communities on protected platforms in the shallow waters of the lake. These indigenous lake communities built on stilts reminded the early Spanish settlers of Venice; hence the name Little Venice or Venezuela was given to the area. From the late sixteenth century, the province of Venezuela was one of six separate colonial administrative units that occupied what is now the nation of Venezuela. In 1777, Venezuela expanded closer to its present boundaries with the creation of the captaincy-general of Venezuela, with Caracas as its capital.

This administrative reform was part of the Bourbon reforms of King Charles III that aimed at strengthening the defense of the colonial empire and raising revenues from it. The reform was an effort of the weakened Spanish rule to ward off British intrusions into the Caribbean. In 1762, Britain briefly occupied Havana, Cuba, and it was able to seize the island of Trinidad in 1797.

The mature colonial society in Venezuela on the eve of independence numbered from about 700,000 to 800,000 people who were rigidly stratified along lines of racially defined castes and legally defined estates. The white elite that lived in the capital, Caracas, included both Spanish-born persons in the colonial administration of military, economic, and religious affairs and native whites (*creoles*). Immediately below them in society were the *pardos*, persons of mixed racial ancestry and of uncertain social standing. The necessary intimacy between master and slave and conqueror and conquered in the frontier conditions of the colony gave rise to the creation of this new person of unique New World existence. In the case of European-Indian unions, they were called *mestizos* (from the Spanish verb *mesclar*, to mix), offspring of European and African unions have been known as *mulattos*.

The growing number of mestizos and mulattos (the pardos) formed the expanding middle sector, or class, of society. The pardos were not slaves or serfs as were their African or Indian parents, nor in most cases could they join the white caste and privileged estates of their other parents; although there is evidence of many being elevated through the love of their European parents and through the purchase of *dispensas de color* (dispensation of color) sold by royal agents. Estimates placed the size of the pardo population at from 40 to 60 percent of the total population around 1800.[1] The elite European community of Spanish and creoles was about 20 percent, while black slaves and a diminishing number of recognizable Indian serfs constituted the remaining 20 percent.

INDEPENDENCE AND THE AGE OF THE CAUDILLO

The conflict between Spanish officials and creoles over economic questions and the willingness of those officials to sponsor the social advancement of able pardos prepared the groundwork for independence. The spreading French and British economic and political penetration into the Caribbean further enhanced the creoles' growing sense of uniqueness from the Spanish. Napoleon's invasion of Spain in 1803 shattered Spanish control at the top, because the keystone of colonial legitimacy was the Spanish monarch. In the eyes of the creole elite, the removal of the Bourbon king from the throne created a vacuum. Between July 5, 1810, the date of Venezuela's declaration of independence, and June 1821, when the last Spanish forces were driven from Venezuela, 11 years of bloodshed and two failures at revolution intervened.

The creole elite fought to create an aristocratic republic to preserve the traditional economic and social system—minus the Spanish officials. Simon Bolívar, who led the independence struggle, feared that the wars of independence would result in a *pardocracia* (pardocracy, rule of the pardos) in Venezuela if too many whites were killed. On the other hand, he feared a racial explosion, like the one in Haiti, if slavery continued.[2] The severity of the independence struggle in Venezuela and the consequent need to mobilize both pardo and black troops destroyed the creole dream. The Spanish forces were defeated eventually, primarily because the *llanero* (plainsmen) cavalry, made up of racially mixed renegade types and under the leadership of José Antonio Páez, sided with the republican forces. In the early years of the struggle for independence, llanero forces had fought ferociously for the royalists and against the hated creoles. The llaneros joined the independence forces after they were convinced that a Spanish victory would mean the maintenance of the rigid, caste-based social structure of the colonial era.

Simon Bolívar attempted to wield together what are today the countries of Colombia, Ecuador, and Venezuela into one nation: Gran Colombia. The strength of local ties and the competitive ambitions of the regional caudillos doomed his enterprise. The fact that Bolívar spent the period from 1821 to 1826 helping to defeat the Spanish forces in Peru prevented him from personally confronting the centrifugal forces that tended to pull Gran Colombia apart. In 1830, Páez led Venezuela's secession from the disintegrating Gran Colombia.

Venezuelan politics between 1830 and 1935 were warlord politics in which regionally based caudillos vied for power "within a network of personal alliances cemented together by community of interest, by force of personality, by ties of friendship and even of family."[3] Politics were concerned with personal gain, not public issues, and were not directed at modifying the basic social and economic structure of the nation. Yet just as the clamor of Adam Smith's marketplace gave rise to an unintended common good, so caudillo politics gave rise to a fluid and open social structure. The rigorous competition among the caudillos was a consequence of their need to involve an increasingly larger share of the population in their campaigns. In addition, the war booty had to be distributed among the victor's following.

The political mobilization that resulted from this style of politics prevented national authority from becoming institutionalized. Possession of national power depended upon naked force and the ability to satisfy the greed of one's supporters. Force ended 13 governments between 1830 and 1935. These 13 successful coups were among 32 major attempts at revolution and 138 minor uprisings recorded during this period.[4] Without a basis for legitimate authority, the nation-state could not operate effectively as the secular arbiter of national life, regulating and directing economic development and social harmony. Similarly, the average Venezuelan was denied means to participate in the affairs of the nation with the result that few residents internalized a sense of civic duty and obligation toward the nation-state.[5] The average Venezuelan's participation in politics amounted to avoiding it, taxation, and conscription into a caudillo struggle.

BUILDING A STABLE NATIONAL AUTHORITY

All the surviving creole power groups of the colonial period (the cocoa, sugar, and coffee plantation owners, the large cattle ranchers, and the commercial bourgeoisie) supported Páez in his creation of an independent Venezuelan state. The constitution of 1830 attempted to ensure the political dominance of the propertied classes through high property and literacy qualifications for voting and for holding office, through indirect elections, and through the preservation of slavery. Under the presidency of Páez, Venezuela passed its first 16 years of independence recuperating from the devastation of the wars of independence.

The devastation of the independence struggle and the fact that power had devolved to the local community and to local elites prevented the reconstruction of the colonial social and economic order. The fragmented republican elite, unlike its closed colonial predecessor, comprised many new men, including mestizo and mulatto caudillos who acquired property from the sequestered royalist lands in payment for their past military service. They, and those who followed them, utilized politics for personal advancement.

A period of anarchy followed the end of Páez's rule in 1846. This period of internecine caudillo conflict forced the mobilization of new groups and the opening up of the social structure. For example, the Monagas brothers, José Tadeo and José Gregorio, abolished slavery while in power in 1854 to weaken their rivals and to gain additional support in time of rebellion. The Federal War (1859–63), which followed the Monagas rule, resulted in 40,000 fatalities and national economic ruin. However, the conflicts brought social equality to Venezuela. According to one Venezuelan historian:

> The Federal War swept away the last forces opposed to social equality since it brought to power representation of the lower ranks of the people, just as they exist still: unable to read and write, with no concept of citizenship, and therefore with no understanding of the exercise of government as the toil of administration and the higher direction of the social group which is the nation.[6]

The period of anarchy associated with the Federal War ended with the rise of national caudillos. Venezuela's strengthened ties with world trade increased the importance of the national capital; there the government leaders worked to extract resources from the international environment in the form of loans and investments, which they used to enhance their coercive power. Increased international trade also enriched the coffers of the government through customs duties.

The reason for this increased flow of wealth into Venezuela due to its reintegration into world trade in the late nineteenth century was simple. As Europe and the United States began to prosper as the result of industrialization and technology, their populations acquired new tastes and needs for Venezuelan products. The first export boom in Venezuela was based on tropical products such as coffee and cocoa.

The first of the national caudillos to undertake the task of nationalizing political power was Antonio Guzmán Blanco, son of the Liberal Party's founder. The period of his dominance (1870–88), the Guzmanía, witnessed the replacement of the self-sufficient *hacienda* (plantation) economy with a profit-oriented, national market system able to respond to Europe's increased demand for tropical agricultural products such as coffee and cocoa. Postal and telegraph service spread to all major cities. Port facilities were improved, and the first railroad in Venezuela was inaugurated.

Guzmán Blanco also moved to end the autonomy of the two most powerful extragovernmental institutions: the Roman Catholic Church and the regional caudillos. Only civil marriages were recognized; public registration of births, marriages, and deaths was initiated; and cemeteries were opened to non-Christians. Convents and monasteries were abolished, and some church land was confiscated.[7] In an attempt to deny local caudillos independent sources of income, the control and taxation of the exploitation of natural resources was nationalized. A fixed annual appropriation to local government, the *Situado Constitucional*, replaced previous sources of local revenue, such as the taxation of the salt mines. In effect, local caudillos became pensionnaires of the national government.[8]

Guzmán Blanco also created a national symbol and myth with which to bind the nation. The memory of Simon Bolívar, as national liberator, was revived. Bolívar's remains were returned to Venezuela and housed in a national pantheon. Equestrian statues of the liberator were erected in the central plazas of all the cities, and the first national currency, inaugurated in 1879, was named the bolívar. The election of the president was opened to direct popular vote through public rallies. An impressive public works program, especially in Caracas, increased the people's identification with their nation.

The rise of the Andean hegemony (1899–1945) ended the decade of anarchy that followed the Guzmanía. The first of the national caudillos from the Andean state of Táchira was Cipriano Castro. His eight-year rule is best remembered for the 1902 naval attack and blockade imposed on Venezuela by a tripartite European unit to force payment of debts owed European claimants. The 27-year dictatorship of Juan Vicente Gómez (1908–35) completed the full nationalization of politics in Venezuela.

The hope of gaining revenue without alienating the powerful landed and commercial interests during his consolidation of power led Gómez, and before him Castro, to grant several concessions to foreign oil prospectors. Revenue from petroleum enabled Gómez to build a modern army with which he crushed the power of the local caudillos. Before the establishment of a national military academy and the creation of a career corps of military officers in the first decade of this century, no real Venezuelan army had existed. The national armed forces had been only a collection of the various caudillo-controlled militias. Gómez's army imposed national integration, but it was essentially the private instrument of his Andean clique: 65 percent of its commanders came from Gómez's home state, Táchira.[9] Young lieutenants would even occasionally be assigned as foremen of common laborers on Gómez's own ranches.

Petroleum dramatically transformed Venezuela. Although substantial oil exportation did not begin until 1921, by 1928 it accounted for more than 90 percent of the nation's foreign exchange. Agricultural production stagnated as peasants flocked to the oil camps and cities to seek employment. Large landowners flocked to the capital to take advantage of the opportunities opening up in government and petroleum administration.

Gómez feared this newly emerging urban Venezuela and reacted strongly against the formation of any private association that might challenge his rule. Even "a nascent Rotary Club in Caracas" was dissolved for this reason.[10] In 1925, he refused to allow construction of a deep-water channel through the sandbar at the Caribbean mouth of Lake Maracaibo. Gómez feared that if the petroleum-rich city of Maracaibo had free access to the sea, with foreign help it and its surrounding region might seek independence. His decision forced petroleum to be barged to the Dutch-controlled island of Curacao where Shell built one of the world's largest refineries.

Gómez's decision also reflected a limited concept of national interest. The foreign petroleum companies were not required to use Venezuelan labor or refine any Venezuelan petroleum. The labor force itself was kept docile. In response to the world depression, Gómez kept the nation's currency, the bolívar, strong, thus destroying the competitive position of the nation's traditional agricultural economy. At the same time, the nation's strong currency increased the propensity to import manufactured goods and made the eventual substitution of national products difficult. Between July 1919 and June 1936, at best 7 percent of the gross earnings of the petroleum companies had been paid to the Venezuelan government.[11]

THE UNCERTAIN PATH TOWARD DEMOCRACY

The death of Gómez on December 17, 1935, marked the beginning of Venezuela's contemporary political development. Between 1936 and the present, occasionally violent, political experimentation has intervened as competing groups have struggled to control the nation's destiny. Four distinct political regimes have ruled the nation since 1936. Each regime has differed radically from its predecessor in political style, development policy, and personnel. Each was initiated in a period of violence and military coup. Each new regime, therefore, promulgated a new constitution to establish both a legal basis and justification for its rule: the constitutions of 1936, 1947, 1953, and 1961.

The politics of the next three decades (1936–64) focused on the clash between two modernizing elites. The first could properly be labeled the modernizing establishment, whereas the second constituted a middle-class counterelite. Although the modernizing establishment was associated with the traditionalist Gómez regime, it aspired to sponsor the social and economic development of the country and the limited and constrained extension of political participation. Venezuela's establishment in 1936 was by no means a rigid caste-based oligarchy. Instead it was an amalgamation of new men: entrepreneurs and top professionals (many of whom were the sons of relatively recent immigrants) and the sons of the Andean followers of Cipriano Castro and Gómez, as well as the traditional landed and urban elites.

Two crucial reasons for the establishment's willingness to sponsor national economic modernization were the shock caused by the world depression and the realization of the nation's dependence on petroleum. Within a year of Wall Street's collapse in 1929, 40 percent of the petroleum workers in Venezuela had been fired, and government revenue had dropped 26 percent.[12]

Opposing the modernizing establishment was an articulate segment of the emerging middle class whose recruitment into politics dated from an anti-Gómez student riot in February 1928 and the attempted revolution in April of that year. The three young students who were responsible for converting student-week festivities into an antigovernment demonstration have all had distinguished political careers: Raúl Leoni, president from 1964 to 1969; Rómulo Betancourt, president from 1959 to 1964 and president of the revolutionary junta from 1945 to 1947; and Jóvito Villalba, the acknowledged winner of the 1952 elections. These men and their associates have received, collectively, the label, the "Generation of '28."[13] This Generation of '28 made its bid for power on the basis of its mobilization and leadership of politically inexperienced workers and peasants.

The Modernizing Establishment (1936–45)

Gómez's death was followed by a period of turmoil. In this turmoil, Minister of War Eleazar López Contreras assumed control of the presidency, having defeated the attempts of the traditionalist Gómecistas to continue the old caudillo-style rule. López Contreras saw himself as Venezuela's guide through the post-Gómez era and aimed at avoiding the extremes of revolution and despotism. Political amnesty was proclaimed, and the exiled student leaders of the Generation of '28 returned to create the nation's first modern political movement, the *Movimiento de Organización Venezolana* (ORVE). This movement was quickly superseded by the *Partido Democrática*

Nacional (PDN) led by Rómulo Betancourt.[14] Supporters of López Contreras formed their own group, *Agrupaciones Cívicas Bolivarianas* (Bolivarian Civic Association).

The López Contreras government laid the foundation for a modern administrative state. The 1936 constitution explicitly recognized the need for governmental guidance and coordination of national development. The government created an industrial development bank, expanded the activities of the Agriculture and Livestock Bank, and created a central bank. In 1938, López Contreras enacted Venezuela's first modern petroleum legislation. The public administration was placed in the hands of competent civilians as the president cautiously moved to separate the civilian from the military aspects of government. Whereas 19 out of the 23 governors were military men in 1936, only four such military state governors remained in 1941.[15]

López's political reforms (of the Burkean-Whig variety) were indirect elections of the national government officials and limited suffrage. Adult, literate males, no more than 15 percent of the adult population in 1936, elected city council members and state assembly representatives. Local legislatures elected the national congress, which, in turn, elected the president.

The government granted the workers the right to organize under the 1936 labor law. However, the PDN leaders quickly organized the petroleum workers and instigated a general strike in December 1936, led by the 20,000 petroleum workers. In reaction, López Contreras outlawed the PDN, dissolved the union, and exiled 47 opposition leaders.

In 1941, the López Contreras-dominated congress elected his minister of war and fellow Táchira native, Gen. Isaías Medina Angarita, president. Thirteen of the 140 deputies voted for author Rómulo Gallegos, the unofficial candidate of the clandestine PDN. In office, Medina Angarita considerably liberalized the political system. The leaders of the PDN were allowed to register their party legally and to hold party rallies openly. Before their organization emerged from its clandestine status, its leaders—the Generation of '28, Rómulo Gallegos, Rómulo Betancourt, Raúl Leoni, Luis B. Prieto Figueroa, and Gonzalo Barrios—changed the party's name to *Acción Democrática* (AD). Medina Angarita eventually organized his supporters into the Partido *Democrática Venezolana* (PDV), and the more conservative members of the establishment grouped around former President López Contreras.

Medina Angarita expanded the economic development programs of his predecessor. Petroleum legislation in 1943 raised the government's share of the industry's profits to about 30 percent. During wartime, the companies were pressured by the U.S. government to reach a satisfactory agreement with the Venezuelans. The nation's first income-tax and social-security legislation was passed. Most significantly, the Medina Angarita government in its last few months in office pushed through an agrarian reform law.

In spite of its tremendous achievements, the Medina Angarita government was overthrown by a group of young military officers, organized into the *Unión Patriótica Military*, in cooperation with the leadership of Acción Democrática. This October 18, 1945, coup and the participation in it of the democratic reformers of AD still stirs heated controversy in Venezuela. For some, the October 1945 coup meant democracy for Venezuela, for others, the end of a decade of gradual and peaceful democratization and the advent of 13 years of alternating demogoguery and dictatorship. Much of contemporary Venezuela's political fragmentation and continuing political crisis stems from this event.

Medina Angarita's reforms were aimed primarily at satisfying the demands of the nation's emerging urban sector. However, the Medina Angarita government responded belatedly to the problems of rural unrest and AD's exploitation of it. Medina Angarita asked his minister of Agriculture, Angel Biaggini, to prepare the draft of an agrarian reform bill that would distribute the sizable government landholdings to landless peasants. Young leftist intellectuals like Carlos Irazábal participated in its formulation. Congress passed the agrarian reform law a short time before the Medina Angarita government was overthrown and only after a heated challenge from the more conservative members.[16]

The ORVE-PDN-AD counterelite that was challenging the modernizing establishment depended greatly upon its organization of the peasantry as a political resource in the power struggle. However, without universal suffrage, this tremendous population element could have no impact. Leaders of AD continue to justify their role in the October 1945 coup on the basis of having established universal adult suffrage and the direct popular election of the president. To the ORVE-PDN-AD counterelite, all reforms and limited extensions of the suffrage undertaken by the modernizing establishment were fraudulent.

Those who consider AD's actions in October 1945 to have been unjustified point out that President Medina Angarita had recognized the popular basis of AD and was willing to agree with its leaders on the nomination and election of an independent coalition candidate. This plan could have led the way to basic constitutional reforms and the establishment of universal suffrage. Such an independent coalition candidate was found in the person of Diogenes Escalante, Venezuela's ambassador to Washington. Escalante became ill, and the coalition agreement had to be cancelled. However, AD leader Rómulo Betancourt had contacted the spokesman for the group of military conspirators, Capt. Marcos Pérez Jiménez, in June 1945, long before Escalante's illness occurred.

Following Escalante's illness, President Medina Angarita supported the nomination for president of his civilian agricultural minister, Angel Biaggini. It could have been that, although the elderly Escalante was acceptable to AD, the younger Biaggini was not. Biaggini was the author of the Medina Angarita government's agrarian reform bill and as president could have eroded AD's peasant support by implementing it.

An impressive list of progressive and leftist intellectuals, such as Jóvito Villalba of Generation of '28 fame and others who were to form the Democratic Republican Union Party (URD), announced their support for Angel Biaggini. One future URD leader, Alirio Ugarte Pelayo, greeted Biaggini at the September 1945 PDV nomination convention as "the minister of agrarian reform." Carlos Irazabel and many other leftist intellectuals such as Juan Bautista Fuenmayor and Rodolfo Quintero, who represented the two competitive communist factions, seemed eager to support Biaggini in order to challenge AD's position in the urban labor union leadership. An elected civilian administration of Biaggini might have very well denied AD the electoral hegemony it was to eventually enjoy as Venezuela's principal populist political party. In participating in the October 18, 1945, coup against Medina, AD's leadership may have been motivated as much by cynical opportunism as well as by a sincere desire to carry out extensive social and political reforms.[17]

Mass-Mobilization Democracy (1945–48)

After the 1945 coup, in which there were an estimated 2,500 casualties, AD and the UPM established a seven-man revolutionary junta under Rómulo Betancourt. The three-year rule of AD (1945–48) has been called the *trienio*. The other members were AD leaders Raúl Leoni, Gonzalo Barrios, and Luis B. Prieto Figueroa, Carlos Delgado Chalband and Mario Vargas representing the UPM, and Edmundo Flores, a civilian independent. Within a year, a popularly elected constituent assembly drafted a new constitution that provided for universal suffrage and lowered the voting age to 18. The AD won 78.8 percent of the total vote cast for the constituent assembly. Rómulo Gallegos, the noted author, was elected president under the provisions of the constitution of 1947. The AD also won 38 of the 42 Senate seats and 83 of the 111 seats in the Chamber of Deputies. Gallegos won 74.4 percent of the presidential vote, and AD won 70.8 percent of the congressional vote.

The mobilization and politicalization of the peasants distinguished this government from its predecessors. The Venezuelan Peasants Federation (FCV) (*Federación Campesina de Venezuela*), under AD leadership, reached 43,000 members by 1948, compared with its estimated 6,000 members in 1945. A revolution took place in rural Venezuela as local power shifted from big landowners to semiliterate peasant union leaders. A national labor confederation, the *Confederación de Trabajadores de Venezuela* (CTV), was also created under AD guidance between 1945 and 1948. More than 300,000 urban workers were organized into 13 AD-led federations.

The AD's dominance of the peasant vote dates from this 1945–48 experience. Former AD leader Domingo Alberto Rangel, who broke with the party in 1960, criticized the agrarian reform carried out during the *trienio*, the AD's three-year reign (1945–48). According to Domingo Alberto Rangel, rather than leading the peasant masses in the conquest of the land, Betancourt merely "created a clientele with which to guarantee AD's electoral predominance." Also according to Alberto Rangel, the agrarian reform law passed in October 1948 was "more conservative than that of Medina Angarita."[18] Whether AD led a self-conscious social class or merely created a rural clientele, it irrevocably altered Venezuelan politics.

The key innovations in development policies were made by the revolutionary junta prior to the inauguration of the popularly elected Gallegos government in February 1948. In December 1946, the junta established a 50-50 arrangement guaranteeing the government at least 50 percent of petroleum industry profits. The architect of AD's development policy was Juan Pablo Pérez Alfonzo, minister of Development. The creation of the *Corporación Venezolana de Formento* (Venezuelan Development Corporation) to help finance Venezuela's transformation into a modern agroindustrial economy through loans to private entrepreneurs and direct investment in state corporations was only an extension of the politics instituted by the López Contreras and Medina Angarita governments. The planning for the creation of a state-owned iron and steel industry and petrochemical industry also began during the trienio.

This period also saw the creation of those other political parties that have remained important actors in contemporary Venezuela: COPEI and the URD. The Social Christian Party (COPEI) started as the *Comité para Organización Política y Elecciones Independientes* (Organized Political Committee for Independent Elections) in 1946. Its founders were Roman Catholic intellectuals who broke with their more secular and leftist colleagues in the Venezuelan Students Federation to form their own National Students Union in the late 1930s. Rafael Caldera has been COPEI's leader since 1946. Between 1945 and 1948, COPEI was a conservative Catholic party with a solid regional base in the tristate Andean region, the home of Venezuela's rulers between 1899 and 1945.

The second major opposition party was the *Uníon Republicana Democrática* (URD), headed by Jóvito Villalba. Villalba had broken with his associates of the Generation of '28 in the days of the PDN and had become quite close to President Medina Angarita. The URD, in the 1945–48 period, was the spokesman for urban middle-class interests seeking a government of unity and calmness rather than the tension-producing, partisan orientation of AD.

The Venezuelan Communist Party (PCV), which was created in 1931, also was able to function publicly in its own name. During the Medina Angarita period, it had operated behind a front organization, the *Unión Popular Venezolana*. Men like Rómulo Betancourt, who later were to organize AD, briefly associated with the PCV in the early 1930s. Their idealistic attraction to communism was quickly dissipated once they became acquainted with its dogmatic rigidity and the awesome degree of Soviet domination associated with it.

In spite of the many accomplishments of the three-year AD rule, the excesses of the regime were alienating the urban middle class, the Catholic Church, and the armed forces faster than the regime could organize and discipline the labor and peasant unions. The AD government held trials for show and exiled men such as Arturo Uslar Pietri and Rodolfo Rojas, whose only crime in many Venezuelans' eyes was their service to deposed President Medina Angarita. The rhetoric of the regime frightened and angered Catholics and many in the urban middle class. The great difficulties the regime faced in translating its rhetoric into reality also led to the erosion of its mass support.

That erosion was evident in the almost 50 percent decrease in voter turnout from the December 1947 presidential and congressional elections to the May 1948 municipal elections. The crucial erosion of AD's support occurred among the army officers who originally helped bring the party into power in 1945. The military was disturbed by the tension-producing AD rhetoric and approached President Gallegos in late 1948, insisting that COPEI be given representation in the government and that AD party president Rómulo Betancourt be exiled. Gallegos refused, and on November 24, 1948, the military overthrew him in a bloodless coup.

In place of the highly politicized development program of the AD regime with its emphasis on human resource development—that is, education and agrarian reform—Pérez Jiménez focused his regime on the "conquest of the physical environment." Pérez Jiménez's program of physical projects was aided by the boom in petroleum production engendered by the Korean conflict and the enforced depoliticalization of policy making. Government revenue doubled between 1951 and 1956. The government projects included luxury hotels and military officers' clubs, and it helped to finance the helicoide, an elaborate, multistoried, commercial center that has never been completed.

Pérez Jiménez then moved to legitimize his own rule through an election in late 1952. He allowed both the URD and the COPEI to contest this election. However, after it became apparent that the URD had won the November 30, 1952, election, it was annulled and many URD leaders were exiled. The COPEI, minus its right wing, assumed a more hostile attitude toward the regime. The increasingly personalist and corrupt nature of the dictatorship alienated a growing percentage of the urban middle class and the armed forces.

In 1954, following the tenth Organization of American States (OAS) conference held in Caracas, which legitimized the CIA overthrow of an elected government in Guatemala, President Eisenhower awarded Marcos Pérez Jiménez the Legion of Merit. According to the letter that accompanied this award, Pérez Jiménez was praised for his sound economic policies, for combating communism, and for social programs.[19] This unfortunate cold-war gesture has occasionally come back to haunt U.S.-Venezuelan relations.

Following the emergence of Pérez Jiménez as strong man, the repression of AD's clandestine organization increased. In 1952, Leonardo Ruiz Pineda, head of the AD underground was killed. In the 1950s, young COPEI party members such as Valmore Acevedo Amaya and José Rodolfo Cardenas, more oriented to COPEI as a social Christian party, were imprisoned by the regime. Even Luis Hererra Campins (COPEI president 1979–84) was briefly imprisoned. According to Donald Herman, AD, COPEI, URD, and PCV party leaders had the opportunity to meet, to develop shared interests against the dictatorship, and to develop a sense of "political cooperation" while in prison.[20]

Democratic Reconciliation (1958–Present)

The petroleum boom of the 1950s brought impressive changes to Venezuela. The rural population, AD's electoral bailiwick in the 1945–48 period, had been reduced to 32.5 percent of the population in 1961, whereas a decade before it was 52.1 percent. In the decade, adult literacy increased to 65.2 percent, and the number of professionals and technicians in the labor force increased to 5.8 percent from 3.5 percent.

The exiled AD leadership under Rómulo Betancourt met in the 1956 convention in Puerto Rico and adopted a new tactic, which de-emphasized the leftist nature of the party. Communications between AD leaders and those of the other parties, URD and COPEI, increased. The leaders of all three political parties—Rómulo Betancourt (AD), Jóvito Villalba (URD), and Rafael Caldera (COPEI)—met in New York in December 1957.[21]

The nation's social and economic establishment played a significant role in the events that led to the overthrow of Pérez Jiménez. Two distinguished Venezuelan businessmen, Eugenio Mendoza and Mario Diez, an executive of the Venezuelan subsidiary of the First National City Bank, also turned up in New York at this time to chat with the exiled political leaders. On January 10, 1958, a number of prominent business and social leaders published a "Declaration Concerning the

National Political Situation," which called for a return of "austere, honest government and civil liberties."

The keystone of the understanding reached between the middle-class political leaders and the social and economic establishment was the latter's expressed willingness to support the return of a mass-based democracy and middle-class political rule via free, competitive elections. In return, the elected political elite would undertake a program of moderate, broad-based reforms, avoiding excessively divisive issues and direct challenges to the prerogatives of such power contenders as the church, the military, and the business community. On January 23, 1958, Pérez Jiménez was overthrown by a broad-based coalition of military officers and businessmen.[22]

Following the fall of Pérez Jiménez in January 1958, two distinguished entrepreneurs, Eugenio Mendoza and Blas Lamberti joined the three military men in the junta. In protest against the inadequate protection offered Vice-president Nixon during his May 1958 visit and the ensuing riots, Mendoza and Lamberti resigned from the provisional government. Two other civilians, Arturo Sosa and Eduardo Sanabria, replaced them. The primary task of the provisional government was to organize the return to representative government. The exiled AD and URD leaders flocked back to Venezuela to rebuild their party organizations.

The return to electoral democracy in 1958 required the cooperation of all parties from COPEI on the right to the PCV on the left. Rightist military factions tried in July and September to prevent the return to open electoral politics. Mass demonstrations sponsored by all the political parties were instrumental in enabling the provisional president, Wolfgang Larrazábal, to defeat these coup attempts. In late October, AD, URD, and COPEI signed the *Punto Fijo* agreement, a pact by which all three parties agreed to form a government of national unity, regardless of the results of the December elections.

The rebuilt AD organization came through with an impressive rural vote to elect party leader Rómulo Betancourt president for 1959–64. Betancourt received 49.2 percent of the votes cast. Faced with a major economic depression and violent challenges from both rightist military factions and leftist revolutionaries, the government of national unity began to come apart in late 1960. In April of that year, the radical, pro-Fidel Castro youth of AD left the party. Violence between the radical left and the government began in October

1960. The URD also left the government coalition at that time. A second split in AD occurred in early 1962, when a group of middle-level party leaders broke with the old-guard leaders.

Aided by a recovery in oil production, AD won the 1963 elections and elected Raúl Leoni president for the 1964–69 term. However, Leoni's plurality was reduced to 32.8 percent of the vote. In order to hold the 1963 elections, the government had first to beat back a determined effort by leftist terrorists to sabotage the electoral process.

The Leoni government (1964–69) was essentially one of consolidation and reconciliation. A concordat was signed with the Vatican and a broad-based government of reconciliation was created, including the URD and a new probusiness party, the National Democratic Front. This coalition government collapsed in 1966. The principal opposition party, COPEI, followed a policy of restraint until the 1966 income-tax-reform controversy. Intense conflict was engendered by this income-tax reform proposal.

In late 1967, AD split for the third time. The issue causing this split was the choice of the party's presidential candidate. One faction insisted upon the party president, Luis Beltrán Prieto Figueroa. The other faction, with support from former President Betancourt, insisted upon Interior Minister Gonzalo Barrios. The Prieto Figueroa faction eventually formed its own party, the People's Electoral Movement. This split in AD enabled COPEI's Rafael Caldera to win the 1968 presidential election with a plurality of 32,000 votes over AD's candidate. Caldera received 29.1 percent of the vote in a four-man race. Both the communists and the supporters of former dictator Pérez Jiménez, organized into the Nationalist Civic Crusade, were allowed to contest the election.

President Caldera's minority government survived in office with little difficulty. The leftist guerrilla operations had all but ended. In the congress, AD, which retained the major block of seats in both houses, has been a responsible loyal opposition as were MEP and the URD. The crucial Petroleum Reversion Law was passed in 1971, and Venezuela joined the Andean Pact in 1973.

In 1973, the party fragmentation of the 1958–68 period ended. AD and COPEI dominated the 1973 elections, which brought AD back into power. Betancourt's Interior Minister Carlos Andrés Pérez was elected president taking all states with the exception of oil-rich Zulia. The oil shock of 1973 brought to Venezuela a petrobonanza of about $50 billion. Creative corruption and massive mismanagement of the petroleum profits prevented Venezuela from buying its way into modernity. Inflation, crime, and urban congestion increased.

In response, COPEI's Luis Hererra Campins was elected president in 1978. Although the Hererra Campins campaign promised a "government of the poor," the reality of Venezuela in the early 1980s has dictated the need for austerity. Unemployment has increased, and the economy has stagnated in the petrobonanza hangover. In 1983, Venezuela devalued its national currency in response to the growing economic crisis. Disillusionment with representative government has increased since 1980.

POLITICAL CULTURE

Political culture refers to the politically relevant values and attitudes current among the nation's population and how those values and attitudes mesh with the assumptions upon which that nation's political institutions rest. Political culture, together with ideological and legalistic considerations, provides the framework within which scholars study the operation of political institutions and forces. When the functional requirements of the formal, constitutional institutions of a nation are not congruent with its value system, one can expect that the performance of the political system will suffer.

THE VALUE SYSTEM

Venezuela's value system is a synthesis of traditional Latin American values and modern values and orientations. Although this synthesis is in a dynamic state, undergoing subtle and continual change, one should not expect its early replacement by an entirely new system more in accord with the smooth operation of representative democracy. Venezuelans have learned to hedge their bets on democracy. Much of their value system comprises traditional, nondemocratic values that have been successfully adapted to the demands of urbanization and industrialization. The often-told story of the father who encouraged one son to become an army officer and a second to go into business but allowed a third to become the local communist-party leader comes to mind. Whether Venezuela's political future were revolution, reform, or reaction, this family would be secure.

One reason Venezuelans hedge their bets on the viability of representative democratic institutions is the confusion resulting from the discontinuity of the nation's political history: since 1936 three military coups and four constitutions have succeeded one another. The violence that racked the nation during the period from 1958 to 1964, when the present constitutional democracy was being established, has further encouraged caution. Many Venezuelans in picking house paints, automobiles, or dresses avoid the officially designated colors of the various parties lest they be identified by neighbors and associates as supporters of given political parties. Many Venezuelans take the added precaution of joining several political parties to avoid being caught supporting the loser.

The values and attitudes of a particular group of society determine a good part of its members' actions by deciding what are desirable ends for action and defining acceptable means for achieving these ends. The following description of Venezuelan values and attitudes is at best a snapshot of a situation so dynamic that it deserves a technicolor motion picture.[23]

Personalism

Venezuelans tend to give loyalty and obedience to the person in command rather than to his institutional position. The notion that abstract general rules can be impartially applied by an impersonal institution is generally regarded with disbelief. Effective decision making in almost every organization is concentrated in its chief, whether he be a government minister, corporate executive, or political-party president. Few important decisions are delegated to subordinates.

Contemporary Venezuelan leaders appear to be locked into the mold established by the traditional caudillo. For example, they must continually deal with their followers face to face. This personification of public institutions and the personal, often private, nature of decision making obscures the boundary between public affairs and private matters. Leaders of ostensibly impersonal institutions become involved in resolving their followers' most intimate, private difficulties. A study of the Venezuelan Peasants Federation shows that 79.9 percent of the total contacts between local peasant leaders and state leaders were face-to-face meetings. Face-to-face meetings were also the most common means of contact (33.1 percent) between local leaders and national leaders of the Peasants Federation, compared with 11.9 percent of contacts made through correspondence.[24]

Particularism

Venezuelans tend to act for the advancement and defense of the interests of their immediate, primary group relationships—kinship, *compadrazgo* (ritual coparenthood), friendship, or patron-clientelism—rather than for some abstract concept of the common good. Relationships among members of these intimate associations are characterized by mutual trust, mutual aid, and courtesy. The world outside of personal relationships tends to be viewed with apprehension and distrust. In this vast, chaotic world, Venezuelans seek out individuals with whom they have the basis for developing personal relationships and, thereby, the right to make special claims for preference and protection. If relatives cannot be found, relatives of relatives will do, or even an acquaintance from the home town or district. Such a complex and extended chain of primary relationships connecting persons of unequal status is patron-clientelism.

Compadrazgo, a social relationship, arises from the Roman Catholic custom of requiring spiritual sponsorship of a child at baptism. A special, kinship-like relationship and reciprocal obligations develop with the parents of the godchild as a result. Compadrazgo's main function has been to enlarge the interpersonal network of contacts an individual can rely on beyond the confines of blood and marriage ties. In Latin America, numerous occasions in addition to baptism have been developed for the naming of honorary sponsors. Some, such as a child's first day in school or the building of a new house, for example, are purely secular.

Also, Latin Americans tend to stress the relationship between *compadres*, cofathers, rather than that between godparents and godchildren. Increasingly, compadrazgo is serving as a means of functionally linking age equals and perhaps social equals, rather than as an instrument of linking unequals in a patron-clientelist relationship. The emergence of corporate ownership, a cash economy, and a mass-based democracy in Venezuela, with consequently greater emotional distances between social statuses, will probably make compadrazgo an even more common means of consolidating mutual support among social equals.

Modernization has reinforced, not reduced, Venezuelan reliance on traditional values and attitudes because of the uncertainty and insecurity inherent in rapid change. In such a transitional period, individuals have found kinship and patron-clientele groups to be emotional havens. Personalism and particularism have thus become the means by which the individual is integrated into the larger society. Personalism and particularism are especially important because of the ambivalence Venezuelans feel toward authority and status.

Power Ambivalence and Paternalism

Traditionally, power was sought in Venezuela as an end in itself, free from moral restraints. Politics were oriented toward winning rather than toward achieving some abstract concept of the common good. Political combat tended to be total; opposition was equated with treachery, victory with domination. Persons in authority were expected to assert themselves aggressively and were not expected to be sensitive to the damage their arbitrary and self-centered policies wrought. The relationship between leader and led resembled the model of the traditional Latin American male-female relationship. The man was expected to be a virile he-man, or macho, insensitive to the "taken" female and her suffering.

Political power is now more and more often perceived as the means to foment general economic development and social justice. Taking and holding power now depends upon popular support. Themes like reconciliation, national unity, and accommodation provide the framework for politics today. The concept of a loyal opposition has been institutionalized. The legacy of the past and the violence and uncertainty of today's politics have made most Venezuelans ambivalent toward formally constituted authority.

In spite of the social mobility possible in Venezuela, its society remains highly stratified, its people extremely status conscious. The terms used to distinguish the upper and middle stratus (*gente buena*) from the lowest social and economic group (*gente mala*, or *los brutos*) connote differences in moral worth as well as in wealth and prestige.

This sense of hierarchy is enhanced by the Spanish language, which retains a distinction between the polite and casual forms of address. Adults dealing with children and individuals of higher socioeconomic status dealing with persons of lower status frequently admonish their partner in conversation not to use the *tu* form because this form of address is reserved for one's social equals. In an argument witnessed by the author that followed a minor collision of a new Mercedes Benz with a beat-up Chevrolet, the driver of the Mercedes Benz insisted that the other cease addressing him with the familiar second-person form and use instead the more formal *usted*.

One consequence of rapid social change has been greater insecurity in interpersonal relationships and increased reliance on primary groups. The lack of a well-defined role and the status system have forced the Venezuelan to become hesitant, cold, and aloof in general society. Traditional Latin courtesy is inoperative in a mass society, as the extreme discomfiture experienced by the driver of the Mercedes Benz illustrates.

Although modernization has destroyed many of the traditional bases of the old social order, such as privileges of birth and caste, it has created its own "disequalizing factors."[25] Evidence that the period of great social fluidity in Venezuela is ending is accumulating. Initially, industrialization created new occupational roles that could not be fitted into the old social order. Many people recruited to fulfill these new tasks acquired their skills through on-the-job training. Now the established members of the developed sector are utilizing their control of such modern institutions as the educational system, labor unions, and professional associations to impede further mobility.

The cumulative effect of these developments is that many low-income Venezuelans have been locked into a permanent under class, or poverty culture. Ascent out of this subculture is becoming more and more difficult because modern, impersonal obstacles—the need to obtain an elementary education before applying for most jobs, for instance—have been placed in the way of future social mobility. Rough-and-ready types willing to raise themselves socially by on-the-job training and hard work are finding this avenue of advancement increasingly closed.

Even the most casual observer of Venezuela may note a definite correlation between an individual's skin color and his social status. A manifestation of Venezuela's latent racism has resulted from the traditional utilization of European immigrants to perform the nation's crucial technical, professional, and managerial tasks, the assumption being that local persons of mixed racial ancestry did not have the ability. Prior to 1958, many of the nation's traditionalist dictators, themselves of mixed racial ancestry, encouraged the immigration of white Europeans and prohibited immigration of nonwhites from Caribbean islands like Trinidad. In the planned industrial city of Cuidad Guayana, Stinchcombe found the potential for ethnic political explosion great. Immigrant groups that had traditionally filled upper- and middle-level positions in local Venezuelan society had become the objects of intense hostility on the part of native-born mestizos and mulattos.[26]

Although no census of racial identity has been undertaken in this century, a generally accepted estimate is that the Venezuelan population is about 70 percent mestizo and mulatto, 20 percent white, and 10 percent black. Indians in isolated communities are less than one percent of the population. (The term *mestizo* comes from the Spanish word *mesclar*, to mix, and is used to describe offspring of Indian and European unions.) The fact that most of the population is of mixed racial heritage should act as a buffer against any potential racial cleavage.

The avalanche of undocumented immigration from the Caribbean and Colombia in the 1970s increased the prospect of an ethnic-racial cleavage in the future of Venezuela's politics. Along with the nostalgic nationalist claims against two-thirds of Guyana, Venezuelans are also becoming concerned that they may become a minority in their own country by the year 2000. In 1980, the government launched an extensive campaign to register foreign workers. It is increasingly difficult to get a visa to Venezuela.

The Jesuit Center (Centro Gumilla) in its Journal of Orientation, *SIC*, published a special series of articles, "Our Racism," in its February 1982 issue. The racist arrogance of protestant missionaries, the sense that racial stereotypes are current in Venezuela, and the concept of *mestizaje* were the subjects of these articles. The claim that Africa's contribution to the making of Venezuela's identity has not been truly appreciated was also made in the *SIC* issues.[27] Venezuelan mestizos and mulattos tend not to enjoy suntans. A Venezuelan host made this comment to me as he pondered why I enjoyed walking and jogging along the beautiful beaches of his hometown.

THE FAMILY

The family is the basic building block of all societies. In a nation like Venezuela, with deep cultural roots in the Hispanic tradition, the primacy of the family continues into adulthood, whereas in the United States primary loyalty shifts from the family during childhood to more impersonal institutions in adulthood. In the United States, the family acts as the base, supporting the individual's commitment to the values and institutions of the larger society. In Venezuela, the individual tends to go forth into the larger society to enhance his family's position, not to seek new loyalties.

The impact of the family in contemporary Venezuela varies with the subculture. The traditional patriarchal family is still common in the peasant sector. As recently as the 1981 census, half the families were formed by common-law unions rather than by legal marriages. In the 1950 census, common-law unions were most common in those states where squatter farms predominated. Although common-law unions have tended to be stable in rural Venezuela, this has not been the case in those states whose agricultural economy has not encouraged settled family life.

The impact of agrarian reform since 1961 has encouraged the spread of legal marriages among the peasants because through settlement the family assumes contractual obligations with regard to land title and agricultural credit.

In comparison to rural families, the family in the marginal urban subculture has assumed a greater importance to the individual, although its stability is more tenuous. An economy that can at best offer the average male *barrio*, or slum, dweller a temporary and/or low-paying job does not support either a stable family or a secure, dominant male image. Frequent desertions and separations have resulted in the emergence of a matriarchal, or mother-centered, family in the barrios, although when the male is present, he heads the family. Approximately half the children born in Venezuela among low-status groups in the early 1960s were born to husbandless mothers.

This condition of economic deprivation and family disintegration has produced some dedicated revolutionaries.[28] However, the average barrio youngster today expresses his anguish through anomic, violent outbursts rather than through planned, principled dissent.

It is in the established urban subculture and among the elite that the traditional extended family is perhaps most functional as a channel of career advancement. Among the economic elite, the family has maintained its importance as a means of linking one elite individual with another. In the military, it was not unusual in the 1960s to find a number of families, often linked by marriage, in control of a service or corps. Members of two related military families were prominent among the leaders of the military uprisings that shook the nation in 1962.

The totality of the impact of the continuing force of family loyalty on the workings of contemporary Venezuelan society is difficult to ascertain. The retention of the family as the prime object of one's loyalty and the use of the extended kinship network as a significant organizer of one's economic and political activities has been defined as familism. Familism as an ideology, operating through extended, intertwined family networks, plays a considerable role in explaining the political loyalties and activities of many Venezuelans.

There are two dimensions of familism: the extent to which the loyalty and primary attention of the public life of Venezuelans of all social statuses is focused on their family and the extent to which kinship networks organized either as political "bloclets" or as economic family groupings dominate political and economic life in Venezuela. While the literature on Venezuela is replete with references to the importance of familism, there is little objective verification of its operations.

The mid-1960s study funded by the Venezuelan Central University Center for the Study of Development (Cendes) and the Ford Foundation of conflict and consensus generated a cross-sectional view of the strength of familism. Of the 24 strategic groups surveyed as to the object of their primary loyalty, only top government officials, union officials, teachers, student leaders, and government executives considered the nation to be the object of their primary loyalty. All other groups, including elected city councilmen, chose the family as the primary object of their loyalty.[29]

A number of studies of the organization of Venezuela's business community identify about a dozen "family entrepreneurial groupings" that bear the name of the founding entrepreneur, the Mendoza Group, the Vollmer Group, the Neuman Group, and so on are said to play a dominant role in the private sector, although the formal representative associations of the business community, the Federation of Chambers of Industry and Commerce (Fedecamaras) and the Pro-Venezuela Association, are said to be largely run by individuals not associated with the powerful family groupings.

The continuing strength and influence of familism in Venezuela may not be as deleterious to the establishment of representative democracy as those scholars who posited the breakdown-of-traditional-society thesis or the need to create a system of impersonal merit recruitment as a prerequisite for democracy may have thought. The operations of familism cushion the impact of adverse social and economic occurrences in a way that mitigates the conflict these events generate. Venezuela in late 1981 had experienced its third year in a row of austerity and economic slowdown, yet its streets were not filled with teeming, angry mobs and beggars. In part, this has been due to the willingness and ability of interlocked, extended family networks to come to the aid of their own. Part-time employment is provided, favors granted, loans made, business contracts allocated and exchanged within a system of family networks. The sharp edges of Venezuela's increased unemployment since the end of the oil boom in 1980 have in a sense been smoothed out by the operations of amoral familism.

Looking back over the events of the turbulent, politically discontinuous generation 1936–57, one may suggest that the fact that individual members of extended family networks often found themselves on opposing sides may have moderated the violent impact of these political clashes. For example, the two Machado brothers, Gustavo and Eduardo, who created the Venezuelan Communist Party in 1931, are related to the Conde Machado family, one of Venezuela's most distinguished and wealthiest. Luis Lander, a former top leader of Democratic Action and proponent of radical democratic reforms, has as a brother Márquez Lander, who was a top executive of Exxon's Venezuelan subsidiary, Creole Oil Co. Juan Pablo Pérez Alfonzo, the architect of Venezuela's oil policy who died in 1979, and retired Gen. Rafael Alfonzo Ravard, president of Petroleum of Venezuela, Inc. (PDVSA) 1976–81, were also close relatives. Pérez Alfonzo fled Venezuela after the 1948 military coup, while Col. Alfonzo Ravard was involved in the key Guayana development project under the 1948–58 military regime.

The conflict-mitigating role of amoral familism on Venezuelan politics is difficult to quantify.[30] There is evidence to suggest that the revolution of October 1945, which brought AD to power, may not have caused as much social polarization as the passions of the moment might have suggested. Many AD leaders in fact were members of the same extended family networks that had prospered during the previous administrations.

This is not to suggest that the events of October 1945 and the coming to power of AD was a pseudoevent. Substantial changes in Venezuela's social structure and political process were in fact introduced and implemented during the conflict-ridden decade 1936–47. What is being suggested is that October 18, 1945, may not have been the break with the past that some saw in it. Previous periods of unrest in Venezuela had, for example, also led to the formation of a fluid society in which the social distinctions separating the very rich and powerful from the very poor and potent largely faded away.

In contemporary Venezuela, there are also examples of family and friendship ties cutting across party labels and ideological barriers. The younger brother of COPEI Party President Luis Herrera Campins, Pablo Herrera Campins, was elected senator from their home state of Portuguesa on the slate of the opposition AD party. Retired Gen. and former Minister of Defense Luis Enrique Rangel Bourgoin, who announced his candidacy for the 1983 presidential election, was considered to be one of two contenders from the military. He leads the right-wing political party National Rescue (*Rescate Nacional,* RN). His brother is Domingo Alberto Rangel, a veteran leader of Venezuela's radical left since 1960. The son of Domingo Alberto Rangel is a youth leader of AD. This young man has two cousins, the sons of Uncle Luis Enrique, who are young army officers. One of Venezuela's largest construction and land development firms is owned by Daniel Camejo Octavio. Pedro Camejo, his son, was the Trotskyist, Social Workers Party candidate for vice-president in the United States in the 1960s. There is nothing like keeping the class conflict within the family.

THE STATUS OF WOMEN

No discussion of the political role of familism would be complete without a mention of the changing legal, economic, and social status of women in contemporary Venezuela. Although the family code has been reformed to recognize the equality of the sexes in civil issues concerning marriage and divorce, this legal emancipation of the woman is much too recent to allow for definitive comment on some issues. The "biggest machista [male-domination] symbol of Venezuelan society," article 473 of the penal code, was only declared unconstitutional in 1980.[31] This article permitted husbands, brothers, and fathers to kill (with impunity) female members of their family caught in the act of adultery. A traditional concept of womanhood and family honor was the basis for article 473.

Venezuela's politics are still largely a man's world. The first woman minister was appointed in 1969 by former President Caldera. President Herrera Campins created the Ministry of State for the Affairs of Women in 1979. Minister of Women's Affairs, Mercedes Pulido de Briceño is encouraging women to organize to improve their job and family statuses. The accusation has been made that her prime motivation is to increase COPEI support among female voters. The recently formed Association of Secretaries does seem to have COPEI leadership.

The fact that women now constitute 50 percent of university students and are becoming more prominent on city councils, state legislatures, and the national legislature should mean that the remaining male-oriented aspects of the civil code will collapse. The first woman governor has been appointed, the critical Central Bank has a woman vice-president, and three of its 16 department heads are women. Even the police and military are now opening up their corps to women recruits.

Unlike the U.S. situation, there is no militant feminism in Venezuela. As in other aspects of its society, Venezuela is attempting a synthesis, over time, of traditional values and advanced ideas of equality. Because "it is difficult to combine tradition and liberation and find a husband," the divorce rate in Venezuela is among the highest in Latin America.[32]

NOTES

1. Estimates of Venezuela's population in 1800 can be found in Robert L. Gilmore, *Caudillism and Militarism in Venezuela, 1830–1910* (Athens: Ohio University Press, 1965), pp. 163–64, and John V. Lombardi, "The Patterns of Venezuela's Past" in *Venezuela: The Democratic Experience*, eds. John D. Martz and David J. Myers (New York: Praeger, 1977), pp. 3–27.

2. See James F. King, "A Royalist View of the Colored Caste in the Venezuelan War of Independence," *Hispanic American Historical Review* 32 (November 1953), and John V. Lombardi, "Los Esclavos en la Legislacíon Republicana de Venezuela," *Boletin Historico* (Caracas: Fundación John Boulton, 1967).

3. Gilmore, *Caudillism and Militarism*, p. 50.

4. See the essay by Juan Liscano, "Aspectos de la Vida Social y Política de Venezuela," in Ediciones de la Presidencia de la Republica, *150 Anos de Vida Republicana (1811–1961)* (Caracas: Presidencia de la Republica, 1963), pp. 191–92.

5. See the excellent discussion of the forces of nationalism in Venezuelan politics in José A. Silva Michelena, "Nationalism in Venezuela," in *A Strategy for Research on Social Policy: The Politics of Change in Venezuela*, vol. 1, eds. Frank Bonilla and José A. Silva Michelena (Cambridge: Massachusetts Institute of Technology Press, 1967).

6. Guillermo Moron, *A History of Venezuela*, 2nd ed. (New York: Harper & Row, 1963), p. 150.

7. For a discussion of the church-state conflict in Venezuela, see John Lloyd Mecham, *Church and State in Latin America*, rev. ed. (Chapel Hill: University of North Carolina, 1966), pp. 105–08.

8. Santiago Gerado Suarez, *Evolución Historica del Situado Constitucional* (Caracas: Editorial Arte, 1965).

9. Moron, *A History of Venezuela*, p. 194.

10. Philip B. Taylor, Jr., *The Venezuelan Golpe de Estado of 1958* (Washington, D.C.: Institute for the Comparative Study of Political Systems, 1968), p. 19.

11. Moron, *A History of Venezuela*, p. 197, and Luis Vallenilla, *Oil: The Making of a New Economic Order, Venezuelan Oil and OPEC* (New York: McGraw-Hill, 1975), pp. 40–41.

12. See Franklin Tugwell, "Petroleum and Public Policy in Venezuela, 1959–1969: A Study of Conflict and Interdependence," (Ph.D. dissertation, Columbia University, 1959), and Edwin Lieuwen, *Petroleum in Venezuela* (New York: Russell and Russell, 1959), for accounts of the impact of the world depression on Venezuela.

13. The origin of the Generation of '28 is discussed in John Martz, *Acción Democrática: Evolution of a Modern Political Party in Venezuela* (Princeton, N.J.: Princeton University Press, 1966).

14. A good description of the López Contreras and Medina Angarita governments appears in Rodolfo Luzardo, *Notas Historico-Economicas: 1928–1963* (Caracas: Editorial Sucre, 1963).

15. Winfield J. Burggraff, "Las Fuerzas Armadas de Venezuela en Transición: El Periodo de López Contreras," *Boletin Historico* (Caracas: Fundación John Boulton, May 1968), pp. 195–96.

16. John D. Powell, "The Politics of Agrarian Reform in Venezuela: History, System and Process," (Ph.D. dissertation, University of Wisconsin-Madison, 1966).

17. Robert J. Alexander, *Rómulo Betancourt and the Transformation of Venezuela* (Rutgers, N.J.: Transaction Books, 1982), pp. 219–220.

18. Domingo Alberto Rangel, *Los Andinos en el Poder* (Merida, Venezuela: Talleres Graficos Universitarios, 1966), pp. 321–22.

19. Donald L. Herman, *Christian Democracy in Venezuela* (Chapel Hill: University of North Carolina Press, 1980), p. 47.

20. Ibid.

21. A version of this New York meeting appears in Domingo Alberto Rangel, *La Revolución de las Fantasias* (Caracas: Ediciones OFIDU, 1966), pp. 43–47.

22. Taylor, *The Venezuelan Golpe de Estado*, p. 53.

23. A comprehensive and comprehensible discussion of traditional Latin American values appears in Henry Wells, *The Modernization of Puerto Rico: A Political Study of Changing Values and Institutions* (Cambridge, Mass.: Harvard University Press, 1969).

24. John D. Powell, "Peasant Society and Clientelist Politics," *American Political Science Review* 64 (June 1970): 423n. Powell's article contains an excellent discussion of patron-clientelist politics in peasant society.

25. An excellent discussion of the possible closing down of social fluidity in Venezuela appears in Lisa R. Peattie's study of squatter settlement in the planned industrial city of Guayana, *The View from the Barrio* (Ann Arbor: University of Michigan Press, 1968).

26. Arthur L. Stinchcombe, "Social Attitudes and Planning in the Guayana" in *Planning Urban Growth and Regional Development: The Experience of the Guayana Program of Venezuela*, ed. Lloyd Rodwin (Cambridge: Massachusetts Institute of Technology Press, 1969), p. 418.

27. "Racismo Nuestro," *SIC: Gentro Gumilla* Año VL. No. 442 (Febrero 1982).

28. See the life history and psychological study of one such individual in Walter H. Slote, "Case Analysis of a Revolutionary," in *A Strategy for Research on Social Policy: The Politics of Change in Venezuela*, vol. 1, eds. Frank Bonilla and J.A. Silva Michelena (Cambridge: Massachusetts Institute of Technology Press, 1967), pp. 241–331.

29. See the presentation of the primary loyalty of Venezuelan strategic groups in David E. Blank, *Politics in Venezuela* (Boston: Little, Brown, 1973), Table VII-1, p. 188.

30. Douglas A. Chalmers developed this concept in his, "Parties and Society in Latin America," (Paper presented at the American Political Science Association annual meeting, Washington, D.C., September 2–7, 1968).

31. Angelina Pollak-Elitz, "Women in Venezuela Today," (Paper presented at the 1982 meeting of the Latin American Studies Association, Washington, D.C., March 18–25, 1982, p. 2).

32. Ibid., p. 8.

3

THE POLITICAL IMPULSE

RULES OF THE ELECTORAL GAME

Contemporary Venezuela has one of the broadest-based electorates in Latin America. Since 1958, the electorate has averaged close to 40 percent of the total population. All Venezuelans 18 years of age and older have the right to vote without regard to literacy, gender, or property. The only limitation on universal adult suffrage is the denial of the right to vote to those subject to prison sentences that carry the loss of political right and to personnel of the armed forces on active duty. The 1961 constitution even allows the national congress to pass a law that would give foreign residents the vote in municipal elections. The constitution states that voting is more than a right; it is a civic duty and therefore obligatory for all citizens.

The apparent achievement of a stable, mass-based representative system in Venezuela is impressive in view of the fact that electoral politics had no significance in Venezuela until 1936 and that universal suffrage was first introduced in 1946. The average nineteenth-century election served only to ratify the results of a violent power struggle. Guzmán Blanco (1870–88), in particular, used public elections to receive public acclamation.

The July 1936 constitution, promulgated during the regime of President López Contreras, called for limited male suffrage and indirect elections for national offices. Literate males 21 years of age and older had the right to elect state assembly and city council representatives. The state assemblies in turn chose the national senate members, and the city councils elected their states' delegation to the Chamber of Deputies. The national congress, in joint session, then elected the president. While the qualifications for voting under the 1936 constitution were more restrictive than those of the Gómez era, the electoral process had for the first time in Venezuelan history assumed an important role in the nation's politics. This opening up of electoral politics was aborted in early 1937 after a series of disruptive strikes and the election to congress of several outspoken leaders of the opposition.

President Medina Angarita allowed opposition political parties to contest local elections in the January 1942 municipal-council and state-assembly elections. The limited size of the electorate participating in these local elections is demonstrated by the fact that only 9,000 votes were cast in Caracas, which then had 400,000 residents. The successful establishment of even this limited suffrage depended on the ability of the opposition, principally AD, to utilize protest politics to avoid electoral fraud. In the 1944 elections, local progovernment forces, aided by the local police, efforts to deny the overwhelming AD electoral victory were frustrated by a petroleum workers' strike and noisy mass demonstrations in the petroleum centers. On orders from President Medina Angarita, the votes were recounted and the AD victory recognized.[1]

After the October 1945 coup that overthrew the Medina Angarita government, universal suffrage was promulgated by the AD-dominated revolutionary junta and later by the 1947 constitution. Venezuelans generally remain divided on the significance of the three-year AD rule (the *trienio*) as a period of exemplary democracy. An electoral commission that included members of all major political parties—AD, URD, COPEI, and the PCV—supervised the three elections held between 1945 and 1948, and a few critics of AD charge that the electoral returns were fraudulently manipulated. But AD's opponents complain that they were handicapped in contesting the elections of 1946, 1947, and 1948 because the official party of the regime, AD, was unable to utilize the entire resources of the government to gain impressive majorities. Jóvito Villalba, leader of the opposition Democratic Republic Union (URD) created in 1946 and a former associate of the AD leaders, called for the creation of a "Government of National Concentration" in which all parties, not just AD, would be represented.[2] Villalba reasoned that elections carried out by such a government would better represent public opinion.

Opponents of AD also insisted, however, that the political climate in most provincial areas was not conducive to free, competitive elections. They charged that the impressive AD majorities were obtained by government manipulation of the unsophisticated and illiterate peasants. The government was accused of using convoys of official vehicles to ferry AD supporters to the polls in some cases. These voters were allegedly well supplied with rum.

After the overthrow of the Pérez Jiménez dictatorship in January 1958, electoral politics and universal suffrage were reestablished. The various political parties agreed to make voting compulsory in order to prevent large voter abstention from being utilized by antidemocratic elements to demonstrate the unpopularity of representative democracy. Although the 1961 constitution allowed separate local elections, the political elite had decided until 1978 to hold one general election every five years. Before 1979, all elected officials, from municipal councilmen to the president, were chosen on the same day. By limiting electoral politics to one year out of every five, the political elite hoped to prevent partisan conflict from keeping the populace in a state of constant agitation and turmoil, which might well have precipitated a military coup.

The role elections play in Venezuela is limited to determining which of the competing political-elite personalities or political party will control the presidency for the next five years. There is little support for the notion that the great issues facing the nation are decided by the electorate on election day. There is very little correlation between party ideology and campaign rhetoric, on the one hand, and the postelection behavior of the elected government, on the other. This is due in part to the desire of all factions of the political elite to avoid deep partisan antagonism and policy cleavages that might threaten the institutionalization of the nation's fledgling democratic system.

The independence of public policy making from the electoral process is also due to the political elites' realization that the interests of crucial foreign- and domestic-power contenders must be consulted on many policy initiatives. They, not a plurality of the voters, provide the resources necessary for a program's successful implementation.

Venezuela's electoral process seeks, on one hand, to guarantee that the effective part of government, the presidency, will be stable.

On the other hand, the electoral process seeks to maximize the representation of all significant political factions in the legislature, whose effective power is limited to moderating, modifying, and blocking presidential proposals. The dual goals of the electoral system are reached by having the president elected by a simple unqualified plurality and the national congress elected by proportional representation.

In 1968, Rafael Caldera of the Social Christian Party (COPEI) was elected president with only 29.09 percent of the votes cast, a plurality of only 31,071 votes, or 0.84 percent of the close to 4 million votes cast. In spite of his small electoral plurality, President Caldera's government was stable. Because of the maintenance of a multiparty system since 1958, no elected president has received a majority of the votes cast.

Seats in the national congress, state legislature, and city councils are apportioned by the d'Hondt system of proportional representation; it allocates seats in the legislative bodies one by one after successive rounds of division of the total votes of all parties. For example, in the first round of division, all party totals are divided by one. The party with the highest vote receives the first seat. In the second round, this party's total vote is divided by two and all others are divided by one. This successive division of the party with the highest total of votes continues until all legislative seats assigned to the state are filled.

In order to illustrate the system, assume that three parties, A, B, and C, are competing for three seats. Party A gained 100 votes; party B, 45 votes; and party C, 20 votes. Party A receives the first seat in the first round of division. It also receives the second seat because its total divided by two is still greater than the 45 votes won by B. However, in the third round, party B picks up the third seat since its total 45 is greater than A's 100 divided by three.

A national quotient system is also applied to allow parties that gained a significant number of votes nationwide a chance to get additional legislative representation when the legislative seats assigned it under the d'Hondt system are less than the party's total national vote divided by the national quotient.

In 1963, for example, the URD's total congressional vote, 497, divided by that year's national quotient, 17, was 29. Since the URD received only 27 deputies according to the d'Hondt system, two additional deputy seats were assigned to it. The figure 17,034 was reached by dividing the total congressional vote cast, 2,861,726 by the fixed number of deputy seats, 168.

Allowing all significant political factions seats in congress and the local legislatures proportional representation has aided a climate of peaceful political coexistence. The system of proportional representation and national quotient gave both the relegalized PCV and the supporters of former dictator Pérez Jiménez, organized into the Crusada Civica Nationalista (CCN, Nationalist Civic Crusade), a stake in maintaining representative democracy. The communists received one senator and five national deputies, and the CCN received five senators and 21 deputies in 1968.

The electorate's limited, passive role is further illustrated by the party ballot system. Voting is done by impersonal party ballot on which the names of the candidates do not appear. The only ballot splitting permissible under the present election law is between the big ballot (*tarjeta grande*), used to elect the president, and the small ballot (*tarjeta pequeña*) used to fill all legislative positions: councilmen (prior to 1979), state assemblymen, and national deputies and senators. The voters have no control over the men selected by the various political parties to be the candidates for the national, state, and local legislative body. The representatives themselves are selected from the closed party list in order of their appearance. These representatives may not even be known by their constituency.

Two Venezuelan political parties, AD and COPEI, in particular, have claimed large active memberships. John Martz, in his study of AD, described a series of requirements—familiarizing oneself with party doctrine, regularly attending meetings, and making periodic contributions—as necessary to remaining a member in good standing of AD. To suggest that more than a handful of people in each state fulfills all these requirements is absurd. The former president of AD, Luis B. Prieto Figueroa, admitted to Martz in the 1960s that the collection of dues is largely symbolic and is in fact not done with any regularity.[3]

FORM AND FUNCTION OF THE PARTY SYSTEM

There is little question that political parties are vital to a healthy democracy. There is, however, almost universal puzzlement over their proper political function. Venezuela's public is not alone in its increasing expression of a sense of disappointment, if not outright disenchantment, with the workings of its political party system. At the heart of this democratic dilemma in Venezuela and elsewhere is the question whether political parties are essentially self-serving electoral machines concerned with winning elections or mass-based coalitions united to win elections in order to translate party principles and proposals into public policy and programs.

In answering this question, one finds oneself inherently entangled with additional questions such as what role party ideology and program play in distinguishing the distinctive Venezuelan parties. Do the political parties differ according to their ideology, organizational structure, and electoral bases? Nowhere is the paradox of Venezuela's democracy more apparent than in its political-party system. Filled with complexity and contradiction, Venezuela's evolving political-party system defies easy definition. Venezuela's party leaders and university scholars are aware of this dilemma and are seriously seeking alternative ways of improving the functioning of the political-party system.

One of the impressive contemporary steps taken in the direction of strengthening the party system's links to society was the June 1980 forum on the "Financing of the Political Parties and Democracy in Venezuela," sponsored by the Institute of Political Studies of the Central University of Venezuela (UCV).[4] In addition to debating a number of scholarly papers, the secretaries general of eight political parties participated in the forum.

One of the paradoxes of the Venezuelan party system is that, on the one hand, the parties are structurally organized and articulated as mass parties with mass membership and a notable degree of programmatic and doctrinal self-definition, while, on the other hand, with regard to their financial lubrication and the apparent motivation of their cadre and leaders, the parties seem to act primarily as means of promoting individual careers and fortunes. The promotion of individual careers is associated with the type of political party frequently labeled as elitist or cadre.

Humberto Njaim of the UCV Institute of Political Studies suggested that this paradoxical split-identity of the parties reflects the reality of Venezuela's oil-based economy. In the late 1970s, government revenue, 70 percent of which came from the oil industry, accounted for almost 40 percent of Venezuela's gross domestic product (GDP).[5] In 1980, the public sector, defined to include state-owned "autonomous institutes" and state-owned enterprises, as well as national and local government, accounted for 43.6 percent of GDP.

Venezuela appears to have one of the more statist political economies of the First and Third Worlds. With agricultural employment rapidly declining and the private industrial sector experiencing a relatively slow rate of growth regarding the generation of new employment, there should be little mystery as to why public sector employment in Venezuela has virtually exploded since the petrodollar boom of the mid-1970s. According to one informed estimate, public sector employment climbed from 350,000 persons in 1973 to 750,000 in 1978, and in 1980, it was estimated to have reached between 900,000 and 1 million.[6]

If Venezuela's principal political parties do appear to operate as patronage machines and career ladders regulating access to the approximately 1 million public-sector jobs, it is not because they have sold out their militants and supporters but because they are indeed responding to the needs and demands of their members and mass constituencies. The political parties in Venezuela are primarily populist, seeking to win elections by providing services and ameliorating the impact of painful change on their electoral constituencies.

Where the line is drawn between populism's amelioration of the pains caused by inequality and hierarchical privilege and democratic reforms more purposive efforts to bring about greater equality is the heart of the paradox of political democracy. A lengthy theoretical debate on comparing and contrasting democratic populism and democratic reform is beyond the scope of this chapter. The writer will, however, succinctly portray the function of Venezuela's party system.

A number of political-attitudinal studies involving low-income and working-class Venezuelan samples tend to confirm the instrumental quality of the populist party system. Patronage considerations (*camburismo*, or securing one's share of the bunch of bananas, or *camburs*, being distributed) rather than party ideology or doctrine was the primary motivation for mass membership in the two preeminent populist parties. In contrast to the United States where the political parties have tended to be pure cadre organizations promoting individual careers and serving largely middle-class interests and values, party membership in Venezuela (AD and COPEI) was highest among the rural and urban poor (43 and 42 percent, respectively) and lowest among the professionals, 28 percent.[7]

Despite the Venezuelan public's appreciation of the instrumental value of its political-party system, there is little doubt that public disenchantment with it is on the ascendency. The charge that the political parties live off and exploit social problems rather than move to resolve them is heard with increased frequency, as is the belief that the political parties are led by frail old men, out of touch with national reality and largely existing for the cambur.

The perception that the party leadership in the early 1980s is out of touch with the contemporary reality is based in part on the fact that the Generation of '28 and the Generation of '36 are only now relinquishing their control over the parties they founded in the 1930s and 1940s. Rómulo Betancourt died in late September 1981. Rafael Caldera, however, is COPEI's presidential candidate in 1983. He was his party's candidate in 1947, 1952, 1958, 1963, and in 1968, when he was elected president. This concentration of real power in the hands of the top leadership is a reflection of Venezuela's caudillo tradition.

Despite the rhetoric of using electoral strength to reclaim the forgotten interior, effective decision making in the political parties is highly concentrated in the capital, Caracas. The parties have been organized downward from the inner circle of top leaders in Caracas to the grass roots. The very concern of the political parties to use state power to sponsor economic decentralization and regional development has itself become a critical factor in the accelerating primacy of the capital of Caracas. The growth of government's economic role since 1958 has been one key factor in the capital's phenomenal growth.

The structure of Venezuela's party system remains quite volatile, and its future direction is uncertain. One key reason for this potential for dramatic change is the rapid transformation of Venezuelan society since 1958. A majority of the eligible voters in the 1983 election were born after 1958. It is not at all certain that these voters have the same ties to the parties that the preceeding generation had. About half the national population in 1981 lived in the 10 largest metropolitan areas.[8] Venezuela since 1958 has been transformed from an essentially rural nation to a nation of big-city dwellers. In 1981, the peasant sector accounted for a mere 20 percent of the labor force.

The 1958 election campaign was essentially contested between party organizations that relied heavily on the ability of local leaders to get out the vote of friends and neighbors. The majority of Venezuela's electorate was still basically illiterate and relied on the face-to-face contact between party leaders and electoral support in the local community, on the one hand, and on the personal, charismatic appeal of national party leaders, such as AD's Rómulo Betancourt, COPEI's Rafael Caldera, and URD's Jóvito Villalba, on the other. The 1978 election campaign was essentially a media event. The political-party personnel were no longer friends and neighbors.

The evidence provided by the 1973 and 1978 general elections indicates that the political parties are playing a diminished role in the management and conduct of elections. Independent financiers and fund raisers and a host of professional pollsters and campaign managers largely replaced the party organization as the prime movers in the 1978 general elections. There is much evidence to suggest that the overall management of the 1978 elections for AD and COPEI was in the hands of hired U.S. media and campaign managers.[9]

The successful 1978 campaign of President Luis Hererra Campins may have been as much the campaign of U.S. media and campaign manager David Garth as it was that of COPEI. The unsuccessful campaign of Luis Piñerúa Ordaz hired the U.S. firms of Joseph Napolitan and George Gaither, who had managed the successful AD campaign in 1973. Although both AD leaders and candidate Piñerua appeared to have been skeptical about the ability of these foreign campaign consultants to appreciate Venezuela's political reality, their advice was critical in the 1978 AD campaign.

The international connection of several of Venezuela's political parties also accounted for both technical and financial assistance. The Christian Democratic Party of the Federal Republic of Germany (FRG) has provided campaign aid to COPEI since 1963.[10] FRG Christian Democratic aid to COPEI was forthcoming in 1978. The Soviet Communist Party has been a strong supporter of the Venezuelan Communist Party (PCV), especially since many youthful PCV leaders left the party in 1970 to protest the Soviet invasion of Czechoslovakia in 1968. The break-away elements of the PCV formed the Movement Toward Socialism (MAS), which has become a Eurocommunist party that follows a Marxist-Gramscian strategy rather than a Leninist one. In 1978, the principal Eurocommunist party, the Communist Party of Italy, gave MAS technical and media assistance.

The financial lubrication of the 1973 and 1978 campaigns reached astronomical sums. In 1973, total campaign expenditures were estimated in excess of $104 million. One Venezuelan observer estimated that AD alone spent $88 million. In 1978, estimates of the total campaign expenditures varied between a low of $100 million to a high of $180 million. The fund-raising job for AD in 1978 was in the hands of a manager of the powerful Vollmer family business group. In the 1960s and 1970s, managers associated with the Mendoza family business group had similarly been associated with COPEI. The Vollmers, Mendozas, and the "12 Apostles" (that is, the big 12 families) were accused of being the principal source of financial support to both AD and COPEI. Since Venezuela lacks a political-campaign disclosure law, the exact role big money has played in lubricating the party system may never be known.

The sense that big money controls party politics has distracted from the popular legitimacy of electoral democracy. There is the beginning of support in Venezuela of having both public and state financing of future elections in order to lessen the feared dominance of big money. According to Henry Wells, postelection grants were made to the various political parties after the 1978 elections "based on their respective percentages of the . . . valid small card [legislative] votes."[11]

VENEZUELA'S BROKERAGE CAPABILITIES

This author and others have suggested that Venezuela's populist political style has a brokerage capability as well as a patronage one.[12] Brokerage politics are more than patronage-centered politics, because they are able to deal with issues of abstract public policy and collective interest, such as the allocation of major public works, and with regional development. Patronage politics focus on individual transactions and particularistic goals and are incapable of dealing with large issues such as regional development or agrarian reform. Brokerage politics, on the other hand, avoid the intense group conflict and ideological orientation of programmatic reform politics, while dealing with issues of collective interests.

With regard to the inherent conflict-ridden nature of politics, brokerage-style populist politics prefer to transfer the focus of the conflict from that of a people problem, focusing on redistribution of wealth and prestige, to one of a resource problem, focusing on the distribution of programs and projects. A neglected region populated by angry peasants would get job-generating public works projects of a community-wide nature rather than a program aiming, for example, to divide the region's large farms into peasant cooperatives. The first policy intends to bring the diverse groups in the hypothetical region together, allying them around common development aspirations, while the latter would tend to divide them along class lines. The competition for development among Venezuela's diverse regions and cities is the ideal arena for brokerage politics.

The 1978 decision to separate municipal elections from the general elections and the holding of the first off-year municipal elections in June 1979 enhanced the structural possibilities of brokerage politics. The fact that a unified leftist slate won an impressive 18.5 percent of the 1979 vote in the first separate municipal election indicates that populist politics on the grass-roots level can also shake up the party system.

Party militancy in Venezuela generally means that an individual has shown up at party headquarters to have his name inscribed on the membership rolls. It was also reported in the press that a number of people detained by the policy in Puerto Cabello after the collapse of a 1962 military uprising in that city had three or four different party-membership booklets on their persons.

The average, nonpolitically involved Venezuelan, whether peasant, barrio dweller, or industrial worker, relates indirectly to the national political-party system as the client of some local leader, or broker. The political brokers act as buffers between locally oriented individuals and the nationally oriented political parties. In a sense, the brokers exchange the patronage and support they receive from the political-party leaders for the votes of their clients. The votes of their clients tend to be given as acknowledgment of their relationship and the individual favors done for them by the broker as well as an expression of support for the party of the broker. In other words, the political-party system depends upon the voter-delivery effectiveness of the local political brokers. It is these brokers who constitute the bulk of the political-party activists in the countryside and in the urban slums.[13] Figure 3.1's depiction of Venezuelan party-activists' patterns will be familiar to those students of the old U.S. city machine politician.

One critical study of the role both political parties and politicians have played in post-1958 Venezuela has suggested that their dilemma is based on their being suspended between their own "cult of rhetoric," under which they aspire toward either a Christian Socialist or Democratic Socialist possibility, and the rather blunt reality of Venezuelan culture.[14] Venezuela's political culture and historical legacy appear to dictate that political leaders build support through particularistic ties and patronage payoffs rather than through societal

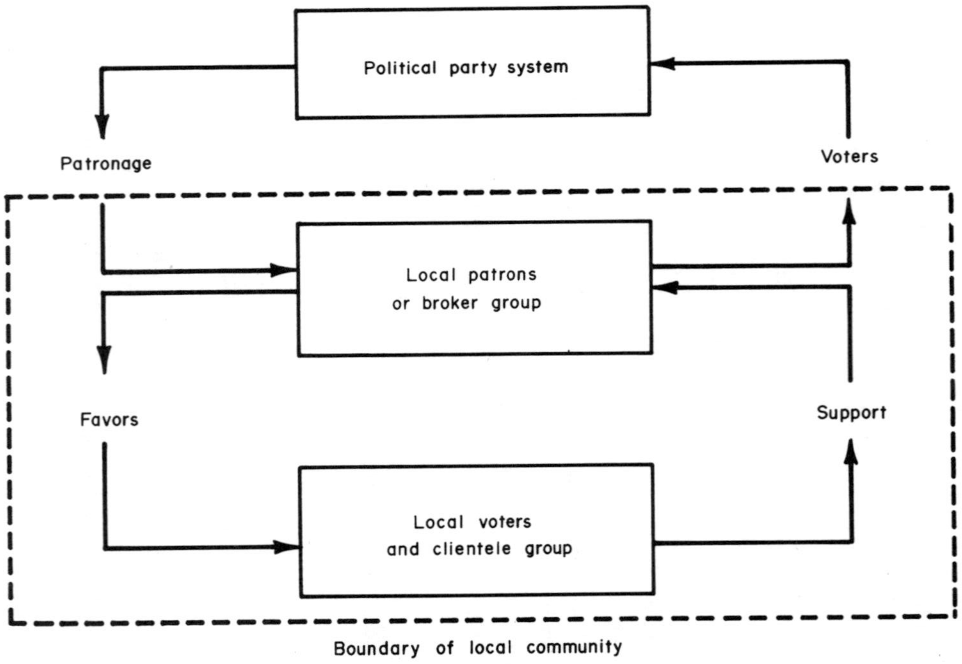

Source: David Eugene Blank, Politics in Venezuela (Boston: Little, Brown and Company, 1973).

Figure 3.1 Two-Step Electoral Exchange System

liberation. The political parties, on the other hand, focus exclusively in their rhetoric and programs on the need to do away with patronage and corruption in the administration of public policies.

There is no easy solution available to the political parties. If they succumb to the requirements of the ongoing culture, they become mere patronage machines and career ladders, thereby losing their ability to inspire and organize public-issue-related constituencies. However, if they exclusively follow their own rhetoric, they may lose their ability to relate to the realities of Venezuela's political culture. This could cost them, and Venezuela's democracy, the ability to survive during a period of economic despair.

NOTES

1. Interestingly, accounts of this incident by the late President Medina Angarita and former President Rómulo Betancourt were quite similar. See Rómulo Betancourt, *Venezuela: Política y Petróleo* (Caracas: Editorial Senderos, 1967), p. 218, and Isaías Medina Angarita, *Cuarto Años de Democracia* (Caracas: Pensamiento Vivo, 1963), pp. 23–24.

2. Boris Bunimov-Parra, *Introduccion a la Sociologia Electoral Venezolana* (Caracas: Editorial Arte, 1968), pp. 45–50.

3. John D. Martz, *Acción Democrática: Evolution of a Modern Political Party in Venezuela* (Princeton, N.J.: Princeton University Press, 1966), p. 216.

4. Juan Carlos Rey, ed. *El Financiamento de las Partidos Políticos y la Democracia en Venezuela.* (Caracas: Instituto de Estudios Políticos, Universidad Central de Venezuela, 1981).

5. Humberto Njaim, "El Financiamiento de la Marquinaria Partidista en Venezuela," in *El Financiament de las Partidos Políticos y la Democracia en Venezuela*, ed. Juan Carlos Rey (Caracas: Instituto de Estudio Políticos, Universidad Central de Venezuela, 1981).

6. Maricio Garcia Araujo, "The Impact of Petrodollars on the Economy and the Public Sector of Venezuela," (Paper presented at the Tenth National Meeting of the Latin American Studies Association, March 4, 1982, Washington, D.C.), p. 11.

7. Enrique A. Baloyra and John D. Martz, "Participation and Efficacy: Aspects of Peasant Involvement in Political Mobilization." *Comparative Politics* 4 (April 1972), 175.

8. David E. Blank, "The Regional Dimension of Venezuelan Politics" in *Venezuela at the Polls: The National Elections of 1978*, ed. Howard R. Penniman (Washington, D.C.: American Enterprise Institute, 1980), pp. 191–217.

9. This thesis is developed by several authors associated with Penniman, *Venezuela at the Polls*, cited in the previous note, also in that volume, see especially David J. Myers, "The Acción Democrática Campaign," pp. 91–132, and Donald J. Herman, "The Christian Democratic Party," pp. 133–53.

10. Myers, "The Acción Democrática Campaign," pp. 106–07, and Herman "The Christian Democratic Party," p. 138. There are several Venezuelan accounts of the 1973 and 1978 elections. See especially Domingo Alberto Rangel, *Elección 1973, El Gran Negocio* (Caracas: Vadell Hermanos, 1974), and José Agustin Silva Michelena and Heinz Rudolf Sonntag, *El Proceso Electoral de 1978: Su Perspectiva Historica Estructural* (Caracas: Ateneo de Caracas, 1979).

11. Henry Wells, "The Conduct of Venezuelan Elections: Rules and Practice" in *Venezuela at the Polls: The National Elections of 1978*, ed. Howard R. Penniman (Washington, D.C.: American Enterprise Institute, 1980), pp. 45–47.

12. Blank, "The Regional Dimension," and Arturo Valenzuela, *Political Brokers in Chile: Local Government in a Centralized Polity* (Durham, N.C.: Duke University Press, 1977).

13. A description of this two-step exchange system appears in Douglas A. Chalmers, "Parties and Society in Latin America," (Paper presented at the annual convention of the American Political Science Association, Washington, D.C., September 2–7, 1968). Also see John D. Powell, "Peasant Society and Clientelist Politics," *American Political Science Review* 64 (June 1970), pp. 411–25.

14. Senta Essenfeld de Breur, *La Cara Oculta del Desarrollo* (Caracas: Ateneo de Caracas, 1982).

4

THE POLITICAL IMPULSE II: THE CONTEMPORARY PARTY SYSTEM

Despite the continuing fluidity of Venezuela's party system, the results of the 1973 and 1978 general elections seem to confirm a tripartite structure of the contemporary party system. The central core of the party system is the increasing domination of the popular vote for president and legislative bodies between the two populist parties that have constituted the post-1958 establishment: AD and COPEI. In 1973 and 1978, these two populist parties gained more than 89 and 88 percent of the presidential vote (see Table 4.1).[1] A third populist party, the Democratic Republican Union (URD), whose 1958 presidential candidate, Wolfgang Larrazábal, gained 35 percent of the vote, has not fared well since that impressive showing and has now been reduced to the status of a minor party.

A number of more radical, electoral-protest parties have emerged since 1958 to the left of the dominant populist establishment. These leftist parties and other radical protest parties and groups that have their principal theater of operations outside the electoral arena constitute the second element of the party system. The third element of the system has been the episodic appearance of antipopulist, independent, and antisystem parties. These antipopulist movements or rightist electoral-protest efforts have generally been one-shot phenomena.

This tripartite classification of Venezuela's political parties into establishment parties, electoral-protest parties, and antisystem par-

Table 4.1. AD and COPEI Vote, National Elections 1958–78 (in percentages)

	Presidential Elections			*Legislative Elections*		
Year	*AD*	*COPEI*	*AD & COPEI*	*AD*	*COPEI*	*AD & COPEI*
1958	49.2	15.2	64.4	49.6	26.8	76.2
1963	32.8	29.1	52.9	32.7	20.8	53.5
1968	28.2	29.1	57.4	25.5	24.0	49.5
1973	46.5	38.8	80.3	42.8	29.1	71.9
1978	43.3	45.2	88.6	39.7	39.7	79.4

Source: Venezuela, Supreme Electoral Council. Republic of Venezuela results for 1958, 1963, 1968, 1973, and 1978 elections.

ties is of greater utility than one based exclusively on a left-right ideological continuum. Ideological considerations are, however, one component of the self-definition of these parties and will be part of the following discussion.

THE ESTABLISHMENT PARTIES

Democratic Action (AD)

Created in 1941, AD remains the primary populist party in Venezuela although it lost the 1978 presidential elections and was severely embarrassed in the 1979 municipal elections, receiving only about 30 percent of the vote. AD has not only demonstrated electoral resilience in surviving three significant divisions and electoral defeats in 1968 and 1978, but it has also initially survived the death of its founder Rómulo Betancourt, in September 1981. Overcoming the loss of the generation of founders and original supporters is the most critical test a political party can face. The fact that Venezuelan opinion polls in 1982 indicated that AD remained the most likely winner of the 1983 general elections and that the party appeared to be reasonably unified under its new leader and 1983 presidential candidate, Jaime Lusinchi, demonstrates that the party has passed this critical test.[2]

The half-century story of the formation and evolution of AD and the career of its founder, Rómulo Betancourt, has been described elsewhere. What follows is a historical outline designed primarily to enable U.S. readers to comprehend the 1980s rather than to refight the issues of the past. A eulogy to Rómulo Betancourt perhaps best describes the half-century trajectory of both the individual and the populist democracy he did much to create.

> Rómulo Betancourt exemplified the political-man of the transition of Venezuela from a nation led by rural caudillos to a nation led by leaders of urban masses. In many ways, Rómulo was a democratic caudillo. The armed peasant groups of the large land owners [of the nineteenth century] have been transformed into the mass following of charismatic populist leaders that embodied in their personal careers the aspirations of the masses in transit from countryside to city. Bread-land-work is the populist concretization of the modernizing democratic dream.
>
> Achieving power [in 1958] and insisting on keeping it, meant for Rómulo Betancourt the necessity to give up the promise and the dream to do that which was possible.[3]

AD's uneasy attempt to balance its promise to substantially reform Venezuela with the realization that in order to survive it probably would have to largely forgo fulfilling this promise led to three formal divisions of the party: those of 1960, 1962, and 1967. In addition, there is the continuing tension between the new party leader, Jaime Lusinchi, and former President Carlos Andrés Pérez, who dominated the party in the 1970s.

The origins of AD's continuing indefinition and tendency to divide can be found in the 1948–58 period. During the military dictatorship, the youthful AD underground led by the late Leonardo Ruiz Pineda (who was killed in 1952) became increasingly estranged from the older, exiled party leaders led by Betancourt. The AD resistance, according to several sources, seemed determined "to build a broad-based movement which would encompass the entire opposition, including the Communist Party."[4] Betancourt, in exile, on the other hand, was reaching out to the Venezuelan business elite, the Christian Democrats (COPEI), anti-Pérez Jiménista elements in the armed forces, and Nelson Rockefeller.

This generational estrangement widened in the 1957–58 period, which led to the January 23, 1958, overthrow of Pérez Jiménez and the broad-based, elite consensus to institutionalize electoral democracy. While Betancourt met in New York City with representatives of the business sector to coordinate the return of "honest government and civil liberties," AD's resistance leaders were at the same time agitating and organizing, along with their more radical colleagues, the street demonstrations and riots that led up to the January 23 fight of the dictator Jiménez. There are many in Venezuela who insist that it was the business sector and not popular unrest that forced Jiménez out.

The return to electoral democracy in 1958 required the cooperation of all political parties and public support by the business sector and Catholic Archbishop Rafael Arias Blanco. Prior to the December 1958 elections, AD's Betancourt met with COPEI's Caldera and URD's Villalba to agree that regardless of the election results, all three major parties would share in the responsibility of a National Unity coalition. In negotiating this tripartite *Punto Fijo* agreement, Betancourt insisted that the Communist Party (PCV) be excluded from the government.

In the December 1958 elections, AD's rural organizations came through with impressive results, enabling Betancourt to be elected president with 49.9 percent of the vote. However, in metropolitan Caracas, which accounted for about 20 percent of the 1958 vote, and in other industrial cities, AD's vote was most disappointing. In Caracas, AD gained a paltry 13 percent of the vote. Adm. Wolfgang Larrazábal, the man who led the January 23 overthrow of Pérez Jiménez and the presidential candidate of the URD, PCV, and a small independent group, gained no less than 68 percent of the capital's vote.[5]

AD, despite its agreement with the business sector, the Catholic Church hierarchy, and the military to manage itself in the spirit of conciliation and compromise, has insisted that it remains the "party of the people." In 1958 at least, it appeared to be merely the electoral organizer of the peasantry. The urban masses had apparently overwhelmingly rejected it in the cities of the booming central region. AD's post-1958 dilemma of indefinition was compounded by the emergence in January 1959 of Cuba's Fidel Castro as a revolutionary alternative.

By April 1969, in part moved by their own seduction by the myth of the Cuban Revolution and their profound sense of frustration with the old-guard leadership of Betancourt, a considerable number of AD's youthful leaders (the generation of 1958) left the party to form their own Revolutionary Left Movement (MIR). The MIR quickly became the "Fidelista" party, advocating the revolutionary road of Fidel Castro.

The divorce within AD, given the climate of the early 1960s, quickly became violent. Accounts of these events differ according to the political predisposition of the chronicler.[6] Regardless of who fired the first shot, Betancourt's old guard or the youthful MIR, the accelerating cycle of opposition violence and government repression was taking its toll on AD. Venezuela's economy had been depressed by falling oil prices and strained relations with the transnational oil companies. Unemployment hovered between 10 and 15 percent. Violence wracked the country as Venezuela was converted into a cold-war battleground. This strained the consensus within AD between those wishing to push for reforms and those demanding caution.

In early 1962, many middle-level party leaders who had to bear the brunt of rank-and-file resentment, the president of the Venezuelan Peasants Federation (FCV) and 13 of the 21 members of the party's National Executive Committee, split from AD to form their own "AD in opposition." This 1962 division was in essence a revolt of much of the party organization against the government faction. The government AD was successful in reducing the AD-in-opposition group to virtual insignificance by the December 1963 general elections. AD went on to win the hotly contested 1963 presidential elections with a reduced plurality of 32 percent of the vote. Once again, the urban voters tended to decisively reject the party.

Raúl Leoni, a former minister of labor who was not Betancourt's first choice for the party's nomination, became Venezuela's president. AD's electoral domination was ended. COPEI, which had been AD's close ally during the Betancourt administration, opted for the role of AD's loyal opposition. President Leoní and AD once again found themselves at a crossroad. One faction, headed by party Secretary Gen. Jesús Angel Paz Galarraga, wanted AD to reclaim its reformist image and to rebuild its mass support, while a second faction, headed by former President Betancourt, perceived the future of the party to be in its gaining adherents among the growing members of the urban middle class.

Raúl Leoni opted for the Betancourt strategy of continuing collaboration with the business community and in winning over the urban middle class. He formed a broad-based governmental coalition with the new probusiness party, headed by Arturo Uslar Pietri, and with Jovito Villalba's URD. Arturo Uslar Pietri was a close friend and adviser to the late President Medina Angarita. Uslar agreed to be the candidate of the antiparty, urban-middle-class based Independents for a National Front (IPFN), which had earlier sought to have AD and COPEI nominate an independent businessman candidate. Uslar succeeded in winning a dominant plurality of the Metropolitan Caracas vote with 42 percent of the vote in 1963.[7] However, AD's continued, albeit declining, strength in the provincial towns and rural areas won a plurality of presidential votes. Each of five presidential candidates won at least 10 percent of the valid votes cast in 1963.

Given the fragmentation of the party system and the growing electoral alienation of the urban middle class, President Leoní may have acted wisely in defusing the IPFN by inviting it to join in a broad-based coalition. However, for many in AD, this meant that the party of the people was to cogovern with the oligarchy. In 1966, AD tried to push forward a basic reform of the tax system with the aim of raising more revenue and making the system more equitable. The intense controversy generated by this proposal shattered the broad-based government coalition and exhausted AD. The government's retreat on the reform proposals triggered the third and nearly fatal division of AD in 1967.

The internal party crisis over whether to push for reform or conciliation became intertwined with that over the party's 1968 presidential nomination. Paz Galarraga, who had been removed as party secretary general in 1966, pushed the candidacy of Luis Beltrán Prieto Figueroa, a venerable member of the Generation of '28 and an advocate of reform. Rómulo Betancourt was equally insistent that Prieto not get the nomination, and he pressured to have Gonzalo Barrios, another venerable leader of the Generation of '28, get the nomination. An effort to have the party's rank and file decide the issue in a primary election was a disaster, and AD literally split down the middle.

Unlike the case of the AD-in-opposition phenomenon in 1963, the Leoni government in 1967–68 was too weak to pulverize the new schism, which quickly named itself the People's Electoral Movement (MEP). MEP's presidential candidate, former AD party president Prieto Figueroa, received 20 percent of the 1968 presidential vote, while AD's Gonzalo Barrios gained more than 27 percent of the officially recognized tally. AD's vote, however, was 32,906 less than that received by COPEI leader Rafael Caldera, who was elected president in his fifth try for the office.

The phrase "officially recognized tally" is used because this writer has been told by respected leaders of COPEI that AD had in fact tried to defeat Caldera by means of a massive electoral fraud. According to these sources, Caldera's true plurality was closer to 70,000. AD had, however, remained Venezuela's dominant party. Although Caldera defeated Barrios for the presidency, AD outpolled COPEI on the impersonal, party legislative ballots, and it was soon able to control Venezuela's congress.

After an initial period of despair following the loss of the presidency, AD rebounded, and with a combination of organizational resistance, skilled legislative leadership, and sheer luck, AD won the 1973 presidential elections. AD's new leader was President-elect Carlos Andrés Pérez. Carlos Andrés Pérez was born in the Andean state of Táchira in 1922. Táchira was the birthplace of the Andean tyrants Cipriano Castro, Juan Vicente Gómez, and Marcos Pérez Jiménez, who had ruled Venezuela for much of this century.

He had been a tough Interior minister under the Betancourt administration and reportedly matched the pro-Castro guerrillas in ruthlessness and disregard for constitutional niceties. As protégé of Rómulo Betancourt, Carlos Andrés Pérez acquired the reputation of being the leader of AD's pragmatic right wing. This vibrant politician was to lead AD back into power in an impressive near-landslide victory that won a plurality in every state but oil-rich Zulia.[8] Pérez's 1973 campaign called for "democracy with energy" and the creation of a "great Venezuela." AD's rhetoric in 1973 was pure social-democratic reformism: promising a "war on poverty" and against "privilege."

Venezuela's populists were not to be constrained in the late 1970s by the fiscal limitations that had frustrated the Betancourt and Leoni administrations. Promising to "manage abundance with the mentality of scarcity," President Pérez, asked for and received special "emergency powers" from the AD-dominated congress in order to manage the economy largely free of popular intrusion and congressional scrutiny.

The Pérez administration created the Venezuelan Investment Fund (FIV, *Fondo de Inversiones de Venezuela*) as a means of freezing up to half of its petroleum income in foreign investments and other long-term investments designed to avoid overheating the economy.[9] Despite the basic soundness of the *IV National Development Plan 1975–1980* and the concept of the FIV and other industrial, agricultural, and regional development funds, managing the oil bonanza overwhelmed Venezuela and AD.[10]

Behind the populist and social-democratic rhetoric of Pérez was the reality that the president was ruling not with AD, but with a group of elite businessmen and bankers.[11] By 1977, AD and Rómulo Betancourt had enough. The party, with Betancourt back in command, successfully revolted aginst the Pérez administration and imposed its candidate, Luis Piñerúa Ordaz, as the party's 1978 presidential candidate. Piñerúa Ordaz started his 1978 presidential campaign by attacking the "12 apostles," those 12 businessmen and bankers upon whom President Pérez had relied to manage his administration.

President Pérez retaliated by informally encouraging one of the 12, dynamic businessman Diego Arias, to launch his own independent presidential campaign using a group he had created in 1973, *Causa Comun* (Common Cause), as his electoral vehicle. Causa Comun supported Andrés Pérez in 1973.[12]

Economic slowdown, runaway inflation, rising crime, and a sense that the oil bonanza of the 1974–78 period had been grossly mismanaged highlighted the 1978 campaign, which was won by COPEI. One of COPEI's principal campaign slogans in 1978 was "Where has all the money gone?" If the AD administrations of Betancourt and Leoní (1959–69) were criticized for being overly cautious and conciliatory, that of Carlos Andrés Pérez (1974–79), perhaps unfairly, was characterized as being corrupt and big-business dominated.[13]

The 1979–82 period was one of deepening world recession and oil glut. The COPEI administration had to manage the postbonanza hangover of inflation, declining oil prices, and a stagnated economy. This reality enabled AD to once again rebound from defeat. Its 1983 presidential candidate, Jaime Lusinchi, was, as of mid-1982, favored to win. Lusinchi, a 54-year-old pediatrician from eastern Venezuela, was the candidate of the Betancourt faction of the party. One of Rómulo Betancourt's final acts before his death in 1981 was to secure the nomination for Lusinchi and to deny it to David Morales Bello, who had been associated with the Pérez faction of the party.

The early 1982 and 1983 campaign statements of presidential candidate Lusinchi resembled in a remarkable way those of Rómulo Betancourt in 1958. There was a promise to fight for the forgotten peasant and to push rural development and the social vindication of the neglected provincial regions. The labor union bureau of the party was a decisive element in securing Lusinchi's nomination, and it called upon AD to more clearly define itself as a socialist, labor-oriented party. In mid-1983, Lusinchi was well ahead of COPEI's Caldera in the polls.

AD is affiliated with the European-based Socialist International (SI) and was to host the 1982 conference of the SI had it not, at the last moment, cancelled the conference in an effort to deny entrance to representatives of the Sandinista government of Nicaragua. "Observer status" in the SI was originally given to that group in 1977. AD also has close ties with the U.S. Democratic Party, and the U.S. media and campaign experts it hired in 1973 and 1978 have largely worked with Democratic Party candidates.

There are many reform-minded personalities in AD. One need only refer to the career of the late Juan Pablo Pérez Alfonzo, Betancourt's minister of Mines, who died in 1979. Pérez Alfonzo is widely regarded as the founder and driving force behind Venezuela's petroleum policy and as such has been called the "father of OPEC." In his later years, he was to caution his fellow Venezuelans about the pitfalls of the 1973–80 oil bonanza, stating that their democracy "could as easily die from overeating as from hunger." He denounced the aimless consumerism and disdain for hard work and cooperation that had resulted from the flood of oil money into the country in the 1970s. He died in 1979, leaving no great fortune to his family and calling for a return to a "human scale" life style in Venezuela. Pérez Alfonzo should emerge as one of the Third World's most profound thinkers and doers and as Venezuela's most revered personality of the twentieth century.[14]

If Pérez Alfonzo represented the promise of AD, Carlos Andrés Pérez has displayed all its contradictions. He is equally at home discussing the prospects of democratic socialism with the Federal Republic of Germany's Willy Brandt and France's François Mitterrand as he is in discussing Miami real-estate investment with some shady characters. Many believe that Pérez was linked with a suspected murder called "the Carmona Case." Carlos Andrés Pérez will probably be remembered in Venezuela for his investments in massive public-works projects. One critique labeled these often poorly built and mismanaged projects to reflect Pérez's "Pharaonic" mentality. One former AD member defined Pérez: "Our Carlism is fundamentally based on pragmatism, cynicism and conventional thought. . . . It is demagogic and volatile being all things to all people, populist, conservative, social democratic, pro-Fidel, pro-big business and leftist."[15]

AD's social-democratic values have focused on creating the instruments of an assertive, nondependent Venezuela and an active government direction of the national economy. Unlike its Western European social-democratic models, it has not had any notable success in creating a reasonably decent welfare state. Public education, public hospitals, public transportation, and public recreation in Venezuela are all a disgrace. National income may have been less equitably distributed in 1982 than it was in 1958.

The Christian Democratic Party (COPEI)

Venezuela's Christian Democratic or Social Christian Party began its existence as the Organized Political Committee for an Independent Election in 1946 in opposition to the then-secular and leftist AD. Its founder, Rafael Caldera, had originally accepted the post of attorney general in the 1945 provisional government, but he quickly resigned to protect alleged AD attempts to repress early Christian Democratic political meetings.

Caldera had been a founder of the Catholic National Students Union in the 1930s and had served in the administration of López Contreras. In the party's first decade of existence (1946–56), it was clearly a conservative Catholic-oriented party. Its electoral base was largely concentrated in the three Andean states. It vehemently opposed AD's alleged Marxism.

The military overthrow of AD in 1948 and the bitter experience of the Pérez Jiménez dictatorship caused the metamorphosis of the party into a more populist Christian Democratic or Social Christian Party of the Western-European mold. In many cases, COPEI leaders received direct help from the German and Italian Christian Democracies in their conversion to democratic populism. A definite leftward movement also followed the reforms initiated in the Roman Catholic Church by the papacy of John XXIII. For reasons of tradition, Venezuela's Christian Democratic Party preserves the COPEI acronym as the label of the party.

In the Betancourt administration 1958–64, Rafael Caldera and COPEI were loyal partners of AD and quickly emerged as Venezuela's second political force in the 1963 elections. The third split in AD, in 1967, brought COPEI to power and Rafael Caldera to the presidency. Caldera's presidency was frustrating for the party. AD remained in control of the congress and the nation's primary political party. Caldera, who had received a plurality of only 29 percent of the vote in 1968, was stymied in attempts to implement his program of "change."

The frustrations of having positions without power and President Caldera's insistence to rule through a multiparty consensus and compromise quite similar to that of the AD administrations (1958–64) led the youthful wing of the party to become increasingly alienated.

One radical Catholic faction left the party to create, with non-COPEI elements, the Christian Left. Two other reform-oriented factions known as the *Astronautas* (astronauts) and *Avanzados* (those who are advanced) attempted to introduce an "anti-imperialist" (to be read as less pro-U.S. and more Third-World oriented) and a radical-reformist orientation to COPEI, with the belief that COPEI should eventually consider moving toward a basic restructuring of Venezuelan society along lines of "communitarianism" (a blend of the best of liberal democracy, Catholic social doctrine, and democratic socialism). President Caldera strongly resisted the Astronauta and Avanzado positions as being out of step with the mainstream of Christian Democracy.[16]

Similar to the problems AD faced in 1968, COPEI was approaching the 1973 elections badly divided. President Caldera's imposition of his choice for the presidential candidate, Interior Minister Lorenzo Fernández, over the party's choice, Luis Hererra Campins, severely weakened the party's 1968 prospects. Caldera's election in 1968 was the first time in Venezuela's history that an opposition group took power through electoral means.

The 1973 elections differed from the three previous ones in that both AD and COPEI just about monopolized the popular vote. Venezuela seemed to be emerging into a two-party system. No other party gained even 6 percent of the legislative ballot vote. As AD stumbled in its management, or more correctly mismanagement, of the petrobillions that flowed into Venezuela between 1974 and 1978 and as the smell of corruption increased, it was natural for the Venezuelan voters to turn to COPEI in 1978.

The party man, Luis Hererra Campins, became COPEI's presidential candidate in 1978. Leaving nothing to chance, Hererra Campins hired a U.S. campaign expert, David Garth, to manage his campaign. Studies have shown that the 1978 COPEI electoral triumph in essence was due not to its own image and program but to the growing disenchantment with AD's Carlos Andrés Pérez.[17] Luis Hererra's campaign hammered at this negative goal by using as its principal campaign slogan the question, "Where has all the money gone?"

In many states, the Hererra Campins candidacy required the votes of the presidential ballots of three other parties as well as that of COPEI in order to defeat AD's Piñerúa Ordaz. Nationwide, however, Hererra's vote on the COPEI presidential ballot would have given him a plurality of 101,382 over Piñerúa Ordaz.

The COPEI 1978 electoral victory was also enhanced by the untimely death of popular television personality Renny Ottolina. Ottolina had announced his presidential candidacy in late 1977 and had created, in emulation of Arturo Uslar Petri's 1963 campaign, a Movement for National Integrity as his campaign vehicle. In March 1978, Ottolina was killed in a crash of a small chartered plane. Had he lived, there was every prospect that his candidacy would have pulled some of the protest vote from Luis Hererra.

The June 1979, off-year municipal elections were an unprecedented triumph for COPEI. In a large part, reflecting the euphoria of the December 1978 electoral triumph and the initial public reaction against Carlos Andrés Pérez, COPEI gained 50 percent of the 1979 municipal-council vote.

The realities of the 1978 campaign required COPEI and Luis Hererra to protect a populist image. The realities of governing Venezuela in the 1979–82 period, a time of a world oil glut and falling oil prices, required the COPEI administration to govern a country suffering a social and economic hangover from the petrodollar debauchery. Luis Hererra's 1979 inaugural address, in which he decried the mortgaging of Venezuela and his failed attempt to bring charges of criminal corruption against former AD President Carlos Andrés Pérez, signaled a new phase in Venezuela's politics.[18] Venezuela's congress eventually "morally sanctioned" Pérez for his involvement in an especially onerous scandal called the "Sierra Nevada" case, but it was quite evident that Venezuelan democracy was not ready to deal harshly with one of its top political leaders in 1980.

COPEI has been forced to govern a stagnated economy plagued by runaway inflation and rising unemployment. Its response has been one of fiscal austerity and a new effort at attracting significant foreign investment. This has led to an unprecedented wave of labor unrest, fomented in part by the fact that AD labor leaders have secured their control over the powerful Venezuelan Confederation of Labor and have not been at all reluctant to use their labor support for political reasons.

While the austere, honest image of Luis Hererra was perceived as a necessary tonic to the spendthrift AD years 1974–79, by 1981, even COPEI leaders were desiring a dynamic president, not an austere one. Luis Hererra's popularity and that of COPEI have plummeted in the polls. As the 1983 elections approached, Luis Hererra, had already lost control of COPEI to former President Rafael Caldera. In 1982,

Rafael Caldera won the party's presidential nomination for 1983 after a bitter struggle with the Hererristas, who supported former Interior Minister Andrés Montes de Oca. Caldera was the party's candidate in 1947, 1952, 1958, 1963, and 1968.

The bitterness of the internal COPEI struggle for the party nomination may in part explain Venezuela's strong support of Argentina in the May 1982 Falklands-Malvinas tragedy. Luis Hererra opted to ride the wave of emotional anti-U.S. feeling due to the war in the South Atlantic in his failed attempt to deny Caldera the nomination. The Hererristas publicly suggested in May 1982 that the United States had betrayed Latin America and that Caldera was tied to the United States.

By late 1982, the COPEI administration appeared to be paralyzed due to the animosity between the Hererristas and Calderistas. Evidence of this lack of government coordination was the 1982 scandal involving the Agricultural Marketing Agency. This gave AD headhunters the opportunity to be used to balance the legacy of the Sierra Nevada scandal. COPEI's prospects for 1983 were also hurt by the emergency of two new antipopulist, middle-class protest parties that directly appealed for defectors from COPEI.

An example of the alleged paralysis of the Hererra administration is the December 1981, "23 of January Housing Project" incident. The "23 of January" is a massive low-income housing project originally built by the Pérez Jiménez dictatorship and renamed for the date he was overthrown. The breakdown of public services and police protection in that project has been frequent. On December 22, 1981, a furious President Hererra demanded of the various public agencies that the project be cleaned up in 48 hours. Press accounts of the project on December 26, 1981 indicated that the presidential order was only partially implemented. The project is located a mere six blocks from the presidential offices, yet an order of the president was not executed regarding the project.

In 1982, Hererra, however, moved dramatically to end Venezuela's economic recession and fiscal mismanagement. The domestic price of gasoline was tripled in order to conserve this resource for export. A dramatic public-sector-budget cut of $2 billion was ordered in April 1982. Hererra also tried to bring the sprawling public-sector state enterprises under some fiscal control.

COPEI has a fundamental problem in relating its day-to-day political action to Catholic moral and social thought. According to one source, "The question is not simply politics versus religion, but rather the way in which religious people see the meaning of their faith in terms of action in the world."[19] In the 1970s, COPEI found itself at odds with radical priests who acted according to the "Liberation Theology." One Belgian priest, who had an activist mission in Caracas's slums, was deported by the Caldera government for "meddling in politics."[20]

The Jesuit Center (*Centro Gumilla*) and its journal of orientation, *SIC*, founded by the Spanish priest Manuel Aguirre Elorriaga S.J. in 1937, played a critical role in the intellectual formation of COPEI's founders, Rafael Caldera, Aristides Calvani, and Lorenzo Fernández. COPEI's generation of 1945, Luis Hererra Campins, Valmore Acevedo, Rodolfo José Cardenas, and Gonzalo Garcia Brustillos, worked closely with Padre Aguire in Caracas's Workers' Circle and other early Christian Democratic groups. In 1981, the Centro Gumilla and *SIC* seemed to have left COPEI to constitute itself as a Catholic faction of the still predominantly Eurocommunist Movement Toward Socialism (MAS).

COPEI and President Hererra Campins seemed to be paying the price of ruling Venezuela at a time of recession and world oil glut. One Hererrista leader, Sen. Valmore Acevedo, perhaps described Hererra's dilemma by stating that the government "refuses to be demagogue regarding the current need for austerity merely to gain in the public opinion polls."[21] Sen. Acevedo organized COPEI's successful 1979 municipal campaign and was a moving force behind the precandidacy of Montes de Oca.

Regardless of the results of the 1983 elections, one can expect both COPEI and AD to remain the two pillars of Venezuela's ongoing populist democracy. Behind the often-intense campaign rhetoric of both parties is a 25-year history of pulling together when the survival of democracy has been threatened. The preservation of electoral democracy and the fomenting of a unique climate of political civility and respect for human rights in Venezuela has been the most honored accomplishment of AD and COPEI.

DEMOCRATIC REPUBLICAN UNION (URD)

Created in 1946 by one of the leaders of the Generation '28, Jóvito Villalba, URD was not expected to survive the 1983 elections or the anticipated retirement of its founder. Both the career of Jóvito Villalba and URD have tended to be ignored by foreign scholars, and the party and its venerable leader have been dismissed as lacking a clear philosophy or as containing only hack politicians.

The URD has not fared well since it left the Betancourt *Punto Fijo* coalition in 1960, and it ceased to be a major populist party in 1973. The reasons for its post-1958 decline are extremely complex and not caused primarily by the party's lack of a coherent philosophy or Villalba's personality. The URD leadership has been much less successful than that of AD and COPEI in preserving an electoral clientele among the peasant and worker organizations. This fact does not demonstrate, however, that it is a qualitatively different party than either AD or COPEI.

Without disparaging the contributions of either Rómulo Betancourt or Rafael Caldera in securing the survival of Venezuela's contemporary democracy, this writer would like to suggest that Jóvito Villalba's role in this endeavor has also been important. The key difference between the URD, on one hand, and the AD and COPEI, on the other, is that Jóvito has performed a most flexible political-broker function, while Betancourt and Caldera converted their parties into the twin pillars of Venezuela's populist establishment.

The key broker role of both Jóvito and the URD is illustrated by the fact that the URD won the 1952 presidential elections against a Pérez Jiménez puppet candidate only to have the dictatorship cancel the election results. Jóvito and the URD were vital in the 1958–60 Punto Fijo tripartite agreement between AD and COPEI. The URD was, once again, the honest broker in arranging to bring the Medinista elements in the Independents for a National Front, organized around Uslar Petri, into a coalition with the populist AD President Raúl Leoni in 1964.

The contribution URD made in defusing the alienation of Venezuela's urban middle-class voters by bringing the Uslar forces into a coalition was vital in securing respect for electoral democracy in the 1964–68 period. Venezuela's last recorded attempt by a military clique to overthrow an elected government occurred in 1966 in the midst of the income-reform crisis. The fact that the URD stayed loyal to AD may have prevented that effort from developing into a serious challenge to democracy.

The cause of the URD's post-1958 decline could possibly be found in its 1960 decision to leave the tripartite Punto Fijo government to protest Betancourt's insistence that Venezuela condemn revolutionary Cuba "without any reservations" at a meeting of Consultation of the Organizations of American States in August 1969. URD Foreign Minister Ignacio Luis Arcayo had attempted to keep the channels of communication opened to Havana, and he resigned his post rather than casting a vote condemning Cuba. The URD supported his action, and Jóvito publicly condemned the San José, Costa Rica, vote as "smacking of the Monroe Doctrine and elements of McCarthyism."[22] Within a few days, the URD left the tripartite coalition and attempted to forge a broad-based loyal democratic opposition. In mid-1960, one could have forecasted that Richard M. Nixon would have been the next U.S. president. Rómulo Betancourt and Rafael Caldera had good reason to batten down the hatches regarding Venezuela's relations with the United States. Vice-president Nixon had perceived that his life was threatened by Caracas mobs during his May 1958 visit to Venezuela, and it would not have made sense to further alienate this politician. Venezuela was suffering from the competition of cheap Middle East oil replacing its markets. The Eisenhower administration had been resisting Venezuela's calls to include it in the program of hemispheric preference granted Mexico and Canada. The year 1960 may not have been an opportune time for Venezuela to play hemispheric honest broker, but Jóvito Villalva and the URD felt otherwise, and they lost.

THE ELECTORAL PROTEST PARTIES

Movement Toward Socialism (MAS)

The origin and trajectory of MAS since its creation in 1970 has been compared to that of AD and its predecessor organizations in the 1936–58 period. Like AD before it, MAS had its origin in a violent attempt to radically transform Venezuela, and like AD, MAS is attempting to become, along with AD and COPEI, one of the pillars of the populist establishment. Similar to AD, the story of MAS is largely the story of one or two politicians: Teodoro and Luben Petkoff and Pompeyo Márquez.

In 1960, the Petkoff brothers and Márquez were leaders of the Venezuelan Communist Party (PCV) youth who felt stymied that the breakaway youth of AD and the left wing of the URD were about to outflank them on the left by becoming the Fidelista group in Venezuela. There was a sense in 1960 that Cuba's experience was the forerunner of a new, Third-World-oriented, revolutionary international force free of the sterile East-West conflict. Career considerations also may have played a role in the decision of the Petkoff brothers and Pompeyo Márquez. Fabricio Ojeda of the URD and Domingo Alberto Rangel and Saez Merida of MIR were threatening to take exclusive control of the radical insurrection. "In fear of being left out of the Revolution [as had happened in Cuba], the PCV also joined the guerrilla movement."[23]

The question of whether the youthful radicals of the PCV, URD, and MIR were forced into violence by government repression or whether they sought from the beginning to violently overthrow the Betancourt government is still debated in Venezuela. Between 1962 and 1967, the future leaders of MAS were among the top leaders of the armed insurrection and clandestine underground.

In 1969, they accepted President Caldera's amnesty decree and returned to open politics. The debacle of their own insurrection attempt, the 1968 Soviet invasion of Czechoslovakia, and the fact that the PCV, which was allowed to contest the 1968 elections as the Union For Progress (UPA), seemed to have no future troubled these individuals. The same people who pestered the PCV into a revolutionary stance in the early 1960s insisted that it end its slavish adherence to the Soviet Union and join Venezuela's reformist mainstream in 1970. This the aged Stalinists of the PCV refused to do. In January 1971, a new party, Movement Toward Socialism, was born with Teodoro Petkoff and Pompeyo Márquez in charge. An estimated 60 percent of the old PCV, including most of its youth and intellectual wings, went with MAS.[24]

MAS has received a friendly press in both Venezuela and the United States. Individuals such as Teodoro Petkoff, who had been feted in Havana in the early 1960s and honored in Moscow, became sought-after participants at U.S.-Latin American conferences in the late 1970s and early 1980s. The New York *Times* published in August 1981 a column on MAS entitled "Venezuelan Surprise, That of Friendly Marxists." Looking much like an Italian movie star, Teodoro Petkoff was featured in a friendly New York *Times* article in November 1980, "Sign of Venezuela's Maturity: Guerrilla in Congress."[25]

What this current trendy approval of MAS does not discuss is the evidence that a very definite financial settlement may have been part of the 1969 amnesty agreement.

Teodoro's brother, Luben Petkoff, was expelled from the PCV in 1966 for maintaining a guerrilla stance in defiance of the "return to the peaceful road" policy voted by the party's political bureau. At that time, the guerrilla's commander, Douglas Bravo, supported the so-called antiparty, proguerrilla line espoused by the late Cuban-Argentine Ernesto "Che" Guevara and by the Frenchman Régis Debray. In 1970, when Luben Petkoff was amnestied and Cuban-Venezuelan relations were on the mend, it appeared that Luben Petkoff received certain commissions for fomenting trade between Cuba and Venezuela. The now-successful business career of Luben Petkoff had been launched.

By 1981, various AD leaders were charging that Luben Petkoff enterprises were receiving favorable treatment by COPEI government officials, perhaps as a means of indirectly lubricating the 1983 MAS campaign in the hope of drawing votes away from AD.[26] Teodoro Petkoff secured his own ascendency in MAS by becoming that party's first president in November 1980. He followed this internal party triumph with an effort to secure MAS hegemony over Venezuela's left. In 1980 and 1981, Teodoro Petkoff was being hailed by the Venezuelan press as the "new Rómulo Betancourt."

In 1973, MAS became Venezuela's third party by gaining 5.29 percent of the legislative vote. In 1978, the party continued to gain by receiving 6.1 percent of the legislative vote. Its share of the June 1979 municipal-elections vote was just shy of 10 percent. Within a decade of its creation, MAS became a critical, balance-of-power electoral element. The ideological bases for the success of MAS is found in its adoption of the philosophy and organizational strategy of the late Italian Communist leader Antonio Gramsci, who died in 1937 while a prisoner of the fascist dictatorship. A critical analysis of Gramsci, published by the Hoover Institution, perhaps inadvertently revealed part of the appeal of Gramsci and MAS in Venezuela: its "retreat to Catholicism's organic community" and compatibility with "Christian messianic expectation and religious belief in a perfected world."[27]

The second reason MAS met with considerable success in the late 1970s is that Venezuela is a youthful country, with a majority of its 1980 population younger than 20 years of age. For the mass of youthful voters in the 18 to 21 category, the issues of the 1930s and 1940s and even the cold-war-related violence of the 1960s are ancient and irrelevant history. MAS is in fact more a party of youth and students than it is of workers and peasants. While it may successfully compete with COPEI in vying for the attention of Venezuela's youthful and activist priests (who are few in number), the remnant of the 1967 MEP break from AD has retained its status as the leftist party with the strongest linkages to the labor movement.

MAS unquestionably passed over the threshold of respectability in 1980. A number of youthful entrepreneurs and businessmen joined the party and were elected to congress on the MAS ticket. The July 1980 cocktail party in honor of Petkoff's election as MAS's first president was held in the exclusive Tamanaco Hotel. Among the hundreds of top businessmen, politicians, cultural leaders, and representatives of the diplomatic corp who greeted Petkoff was the U.S. ambassador. Finally, an additional element in the democratic convergence of MAS and in its gaining electoral respectability has been the activity in Venezuela of the Friedrich Ebert Foundation, affiliated with the Federal Republic of Germany's Social Democratic Party (SPD).

The SPD Friedrich Ebert Foundation has vastly increased its activity in the Third World since Willy Brandt became president of Socialist International and a leader of the North-South Dialogue in 1976. Both Brandt and French Socialist President François Mitterrand have used the international ties of their respective parties as an instrument of projecting French and German interests in Latin America. The Friedrich Ebert people this writer interviewed in 1980 and 1981 clearly considered MAS to be an integral part of Venezuela's "progressive democratic stream."[28]

In 1973 and 1978, the MAS presidential candidate was the independent, former URD member José Vicente Rangel. Rangel had never been a guerrilla and never formally joined MAS. According to John Martz, Rangel was nominated as the MAS presidential candidate in 1978 primarily as a means of preventing a split between Teodoro Petkoff and Pompeyo Márquez.[29] Petkoff was not totally pleased with the arrangement and considered forming his own party. MAS legislative vote was larger than Rangel's presidential vote.

By late 1980, Petkoff was sufficiently entrenched as the MAS leader to have the party formally decide to select its 1983 candidate from among its own members. Rangel quickly retaliated by forming his own New Alternative Electoral Coalition. By late 1981, the PCV and an assortment of other leftist parties announced their support of Rangel and their decision to prevent Petkoff from gaining hegemony over the left. Rangel insisted that his New Alternative was "centrist."

MAS had two other obstacles to surmount in the 1983 elections. Not only did it need to overcome the violent past of its founders, but also it needed to convince many skeptical Venezuelans that its democratic conversion was irreversible. There are those in Venezuela who insist that MAS's peaceful turn is merely tactical.[30]

In August 1982, a faction of MAS led by party Deputy Secretary Gen. Tirso Pinto came out in support of Rangel and against the "personalist dictatorship" of Teodoro Petkoff. The Pinto defection also attacked MAS's convergence with the establishment.

People's Electoral Movement (MEP)

MEP, the result of the most significant division of AD, in 1967, survived the 1968 elections in rather good order. Its notable strength in the labor movement (CTV) was highlighted by the fact that longtime CTV President González Navarro was a MEP leader. González Navarro subsequently returned to AD. AD had lost control over the presidency in 1968 and could not use the economic and repressive power of the state to grind down MEP as it had done in the previous divisions, of 1960 and 1962.

There was, however, a basic weakness in MEP. The essential reason for the party's coming into being was Rómulo Betancourt's veto of Luis Beltrán Prieto Figueroa as the party's 1968 presidential candidate. Betancourt's reasoning was that Prieto would have rekindled all the fears of the 1945–48 AD experience. In addition, Betancourt insisted that Prieto's nomination would have given the AD organization over to Jesús Angel Paz Galarraga, who Betancourt insisted retained his youthful proclivity for Marxism–Leninism.[31]

Prieto and Paz reacted with equal intemperance. They charged that the ability of "Mr. Betancourt" to override the decision of a vast majority of party members in favor of Prieto proved that AD had exhausted its historical mission and was a mere reflection of the ego of this conservative caudillo. Prieto was to take a special umbrage at what he alleged was Betancourt's racism. Prieto was Venezuela's first top politician to clearly display an African heritage. Despite official denials, there is much latent racism in Venezuelan society.[32] Prieto was to include in this 1968 campaign the insistence that he was the "colored" candidate.

MEP leaders asserted in 1967–68 that they represented the democratic-socialist-reform impulse in Venezuela. In the 1970–71 period during which COPEI was adjusting to the fact that it was the government party and AD was sulking in isolation, MEP seemed to be riding high. In Chile, a broad-based democratic-socialist-reform coalition had narrowly won the 1970 elections. MEP more and more evolved toward an uncompromising democratic-socialist-reformist stance. In emulation of the Chilean experience, MEP leaders tried to form a coalition with the URD and FDP (created to support Wolfgang Larrazábal's 1963 campaign). This effort to form a third force in Venezuela's politics failed. Part of MEP's problem was that it moved to consider both the PCV and MAS as potential members of the "new force" electoral coalition it was organizing for 1973. This proved too much for many of its leaders and members; return to AD in the 1972–73 period was impressive.

In 1971, COPEI and AD agreed to reconstitute their working agreement and began to discuss then-common "areas of coincidence" in order, according to Tugwell, to "restore public [and military] confidence in the democratic process."[33] Preventing the successful creation of a third-force coalition led by MEP was one area of coincidence. Joining forces to dislodge MEP labor leaders from key positions in the Confederation of Labor was the top "AD-COPEI coincidence," which was achieved.

Paz Galarraga's efforts in 1973 to convert the promised third-force coalition of MEP and the URD into a new force that included the PCV was a disaster. MEP's vote plunged to a mere 5 percent of the total in the 1973 elections. In 1978, MEP barely received the necessary 1 percent required to maintain its permanent registry as a party. Bitterness has been one force driving MEP's leadership. Prieto, at age 76, ran for president in 1978, largely, according to one top AD source, as a "kamikaze pilot" out to destroy AD's chances of winning.

In 1981, MEP refused to side with either MAS's Petkoff or José Vicente Rangel's "New Alternative." Instead, it nominated Salom Mesa Espinoza, a one-time friend of Rómulo Betancourt, who had been imprisoned by President Pérez because of his alleged involvement in the bizarre kidnapping of a U.S. businessman. In August

1982, MEP's national leadership opted to support the candidacy of José Vicente Rangel.

THE NEW ALTERNATIVE

In November 1980, José Vicente Rangel, the independent, former leftist member of the URD and presidential candidate of MAS in 1973 and 1978, announced the formation of a "New Alternative" electoral coalition. Rangel had just been denied the prospect of becoming MAS's 1983 presidential candidate, and the creation of the New Alternative was his reply. The primary goal of the New Alternative was to organize a broad-based coalition of leftist and "centrist" parties able to beat MAS and Petkoff in the multiparty primary elections that had been tentatively scheduled for late 1982.

Rangel has publicly stated that he sees himself as the Venezuelan version of François Mitterrand, and in his new role as "mature," "centrist," presidential candidate, he refused to be provoked on a number of issues. Rangel has indicated that he was inclined to accept the military's judgment on the purchase of 18 F-16 advanced fighter aircraft and was quite confident at the news that the Association of Retired Senior Military Officers was considering supporting a military candidate.[34] Rangel cited the historical cases in which military men made excellent presidents: DeGaulle in France and Lázaro Cárdenas in Mexico. He also praised López Contreras and Medina Angarita for being decent presidents.[35]

In 1980, largely because of his previous exposure as a presidential candidate and the greater name recognition that afforded him, Rangel outpolled Petkoff in several public opinion polls. However, the 1981 polls indicated that Petkoff was gaining in name recognition and support.[36] Rangel's primary problem is that he is attempting to wield together a most diverse and disparate coalition. Because of this, Rangel seems to be wavering on the issues. Whereas MAS and Petkoff condemned without equivocation the Soviet invasion of Afghanistan, Rangel merely indicated that he opposed in principle all kinds of foreign intervention.

The heterogeneous nature of the New Alternative Coalition, as of late 1982, is subsequently indicated. It should be noted that historically such efforts at broad-based electoral coalitions in Venezuela have not been overly successful. In November 1982, the following groups in addition to MEP publicly proclaimed their support of Rangel.

Venezuelan Communist Party (PCV)

Hector Mújica, the PCV 1978 presidential candidate is a youthful, personable university professor. Mújica has taken over the limelight within the PCV from its aged Stalinist leader, Gustavo Machado. The PCV has concentrated its fury on attacking Teodoro Petkoff and his Gramscian "new socialism" calling it a "fantasy" that serves imperialism (meaning the United States). In May 1981, MAS held an international conference around the theme, "From the Existing Socialism to a New Socialism." In July 1981, the PCV retaliated with its own conference on "True Socialism." For the PCV, following the example of Lenin is the only way of achieving socialism.[37]

According to a PCV article published in the March 1981 Moscow *World Marxist Review*, the party, due to its "ideological strength," is going to be a component of a prospective leftist unity in Venezuela that will confront U.S. imperialism and Teodoro Petkoff's fantasies.[38] The PCV is Venezuela's oldest political party, and the Soviet Communist Party was represented at its 50th anniversary celebration in 1981.

Vanguard Party (VG)

The VG split from the PCV in 1974. Eduardo Machado, brother of PCV President Gustavo Machado. Federico Brito Figueroa and Guillermo Garcia Ponce are its leaders. The party failed to maintain its status as a registered national party in 1978 and 1979.

The Vanguard Party, despite its relatively small size, publishes a quality magazine, *Decada 80*. Its May 1980 (May Day) issue had an article, "China: A Zigzag in Search of Socialism," suggesting that the VG had a friendlier attitude toward China than did the PCV.[39]

Revolutionary Left Movement (MIR)

MIR is the remnant of the 1960 AD division. MIR was restored to legal status as a political party in 1977, having been outlawed in

1963 as a guerrilla group. Perhaps because of the revolutionary mystique of its former guerrilla leaders, MIR emerged as Venezuela's fourth largest party in the 1979 municipal elections, with 2.3 percent of the vote. It ran its own presidential candidate in 1979, Américo Martín.

In 1979, Martín and party Secretary Gen. Moisés Moleiro split. According to one source, the MIR split was due to the problems of disillusionment these former guerrilla fighters found in the electoral arena, especially when there were other leftist competitors. However, David Myers suggested that the real cause of the 1979 split was the squabbling between the rival MIR factions for the choice university-administrative and student-leadership jobs. MIR won impressive victories in key university elections. Apparently the former guerrilla fighters may not have been above a little self-serving corruption.[40]

The faction controlled by Moisés Moleiro has joined forces with MAS. The Martín faction has vacilated between cooperating with the New Alternative and considering an independent campaign. Apparently, Americo Martín and Petkoff are old rivals. They did not get along as competing student and guerrilla leaders in the 1960s.

Socialist League (LS)

This party was organized in 1973 in a division of the OR, one of the guerrilla bands (Organizations of Revolutionaries). It was allegedly responsible for the kidnapping of U.S. businessman William Niehaus. David Nieves, who was elected to congress in 1978 as the LS sole representative (based on the national quotient) was released from prison because of his congressional status.

LS has been called Trotskyist in its ideology. A 1980 "Declaration for a New Alternative" (not to be confused with Rangel's New Alternative) explicitly rejected exclusive "parliamentary involvement with the ongoing system. . . . parliamentary maneuvering is only part of the struggle with the enemy. . . . the principle theatre of operations . . . should be the poor neighborhoods, factories and peasant communities."[41] One may interpret the declaration to mean that the LS involves itself in extraparliamentary, community organization and mobilization activities. Some insist that it is a veiled promise to maintain a guerrilla effort or "people's war." LS Secretary Gen. Julia Escalona was granted amnesty only in 1979 by President Hererra Campins. The LS supported Rangel in 1978, but its share of the legislative vote did not assure it permanent status as a national party. In 1981, LS announced its support of Rangel.

The LS, as well as other radical groups, may have had Libyan financial support. David Nieves was part of the receiving party that greeted Shaati Asmad, an adviser to Libya's Col. Muammar Al Qaddafi, in November 1981. Asmad was the Libyan representative at a multimillion-dollar conference on the "Third Universal Theory" held at Caracas's Hilton Hotel.

The People Advance (El Pueblo Avanca, EPA) and Revolutionary Action Group (GAR)

Both EPA and GAR have Catholic derivations. EPA, led by Edwin Zambrano, comes from a radical faction of students and faculty associated with the Andrés Bello Catholic University. Rafael Iribarren of GAR was a member of the COPEI party. The Christian Left he created split over the issues involved in merging Marxist and Christian values. EPA and GAR supported Rangel.[42]

Radical Cause R and the Revolutionary Rupture Party (PRV)

Two other radical groups did not declare their presidential intent for 1983 by May 1982. Radical Cause (*Causa R*), led by Alfredo Maneiro, has targeted its activities in two working-class communities where it has had considerable success in its grass-roots mobilization. Causa R believes that the job of the left in Venezuela is to build a grass-roots popular movement that serves the people, not the ambitions of the competing left-wing caudillos.

Douglas Bravo is a former PCV guerrilla leader who was expelled from the party in 1967 when he supported the assassination of Julio Iribarren Borges, brother of Venezuela's foreign minister. Cuba's Fidel Castro and Venezuela's PCV were to get involved in a bitter polemic because of this 1967 terrorist assassination. The PCV was insisting on a return to the peaceful road, while Castro was involved in projecting himself as a potential leader of the Third World revolution. Bravo supported Castro's position. In 1970, when Cuban and Venezuelan relations were on the mend, Castro stopped helping the Venezuelan guerrillas. Bravo broke with Castro, insisting that Cuba was only interested in "its own material well-being" and was abandoning the "Continental Revolution."[43]

Bravo was to remain a guerrilla leader in his home region of Falcon and Yaracuy states until accepting President Hererra's amnesty in November 1979. His return to public life in his home state of Falcon was the occasion of a media event and a rather large fiesta. The Bravo family is a traditional landowning family in the state where Douglas was a guerrilla leader.

Bravo has stayed out of the Petkoff-Rangel feud. His political party, organized by associates and family members while he was still operating a clandestine movement, has been called Venezuelan Revolutionary Party–Rupture (calling for a revolutionary rupture). The PRV–Rupture was created in 1979 when Bravo and his former associates split. A July 1981 newspaper account of the retired military general's interest in politics contained the following:

> Despite the fact that the general [retired general Arnaldo Castro Hurtado] has acted with a distinct rightist tendency, paradoxically he has received the support of one of the most famous guerrillas . . . , Douglas Bravo, who has stated publicly that he would have no objections in supporting [the general] if his nationalist conceptualizations are sincere.[44]

EFFORTS TOWARD LEFTIST UNITY

Based in part on the success of the unified leftist state in June 1979 municipal elections, a most serious effort started in 1981 to achieve a unified leftist presidential candidate for the 1983 elections. In 1981, eight political groups formed the National Coordinator of the Left Commission, empowered to bring about this unity either by means of negotiations or by Venezuela's first multiparty primary elections. By late 1982, the competition between Teodoro Petkoff of MAS and Jose Vicente Rangel of the New Alternative doomed the effort.

The Causa R's Alfredo Maneiro came up with a novel third alternative in early 1982. Maneiro proposed Jorge Olavarría, the editor of *Resúmen* magazine as the Unity candidate. Jorge Olavarría was briefly exiled by the Carlos Andrés Pérez administration in late 1978 for investigative journalism that the president found to be too strong. His potential candidacy as the nonpartisan head of a center-left electoral coalition indicates the continuing pressure for change in Venezuela's party system.

The complex nature of the relationship between the personal ambitions of individual top politicians and the evolution of the political-party system is further illustrated by 1981 and 1982 indications that AD's Carlos Andrés Pérez and MAS's Teodoro Petkoff may be heading for rapprochement. Despite the death of Rómulo Betancourt, former President Pérez had lost his control over the AD party organization. Petkoff may threaten to quit MAS if Rangel eventually gets the left's Unity nomination. In 1981, there were signs that Carlos Andrés Pérez and Teodoro Petkoff were considering a potential synchronization of their political efforts.

ANTIPARTY, NONPOPULIST ELECTORAL PROTESTS

Given the populist reality of Venezuela and the consensus in favor of universal suffrage, urban middle-class voter and business groups felt a deep sense of rightlessness because the rural-based populist parties swamped their interests in the electoral process. The sense of rightlessness felt by the urban middle class was heightened by the fact that many of them were not *criollo* (native-stock Venezuelans). According to a 1963 study of strategic groups, 60 percent of the industrial executives and 44 percent of the commercial entrepreneurs were foreign born.[45]

Since 1963, a number of antiparty and nonpopulist electoral protests have emerged to challenge the populist and radical parties. The Independents for a National Front (IPFN) in 1963 formed the first of these electoral phenomena. The creation of the National Civic Crusade (CCN) in 1968 was an even more serious challenge. The CCN was the political arm of those who sought the return to power of former dictator Marcos Pérez Jiménez. The CCN challenge in 1968 was a challenge to the future of the multiparty electoral democracy.

The IPFN ran the distinguished author Arturo Uslar Pietri for president in 1963. The origin of the 1963 antiparty independent movement that became the IPFN was the politicalization of the nation's social and economic elite in 1962 in reaction to the apparent shattering of President Betancourt's political support. In late 1962, representatives of the business elite formed the Association of Independent Venezuelans (AVI). The initial aim of AVI was to use the financial power of the business community to convince both AD and COPEI to nominate jointly an independent presidential candidate whose program would be independent of both parties' populist ideologies.

The politicalization of the business community, however, engendered considerable antiparty sentiment and momentum to launch an independent political movement. While one faction of AVI proceeded to negotiate with AD and COPEI, another distrusted both AD and COPEI sufficiently to pressure Arturo Uslar Pietri, friend and adviser to the late President Medina Angarita, to run for president. In May 1963, Uslar Pietri agreed to head the ticket of the newly formed IPFN.

The IPFN challenge to the established parties became more serious when the former AD president of the FCV, Ramón Quijada, announced his support of Uslar Pietri. Quijada had been a leader of the movement that broke with AD in early 1962. It became clear by early 1963 that the support of the AD-in-opposition groups had all but evaporated. Practical considerations, therefore, induced Quijada to lead AD-in-opposition to join the IPFN, promising to deliver 500,000 peasant votes to Uslar Pietri.

Uslar Pietri apparently dominated his opponents in the use of television. Former AD Secretary Gen. Jesús Angel Paz Galarraga (now a MEP leader) later wrote that Uslar Pietri's electoral impact "coincided with the area covered by TV" regardless of the social class or ideology of the viewer.[46] The IPFN vote was in fact largely limited to urban areas and the central region. Quijada's boast that he could deliver peasant votes proved false. Nevertheless, the close to 470,000 votes Uslar Pietri received in 1963 were most impressive. Five senators and 22 deputies were elected on the IPFN legislative ballot.

Following the elections, Uslar Pietri reorganized the IPFN into a political party, *Frente Nacional Democrático*. Not all the congressmen elected on the antiparty IPFN ballot joined Uslar Pietri in this effort. Even fewer supported him when, in November 1964, he agreed to form a "broad-based government of mutual understanding and reconciliation" with AD and URD. To many IPFN voters, Uslar's joining the AD was a betrayal of his entire electoral campaign.

In 1978, popular television personality Renny Ottolina decided to follow in the footsteps of Uslar Pietri. He created, with the help of other antiparty independents, the Movement For National Integrity (MIN) in 1977. Substantial financial support quickly converted MIN into a serious electoral operation. Renny began his 1978 campaign focusing, in Martz's words, on a single unfailing message, "The country no longer believed in political parties; it had tested them for twenty years and found them wanting."[47] Unlike Uslar's 1963 campaign, that of Renny ended in a tragic plane crash in March 1978. His widow received 30,000 votes in the federal district, and MIN's 85,000 legislative votes nationwide was an impressive memorial to this individual.

Diego Arria, a close collaborator of AD's Carlos Andrés Pérez, launched his own "independent" candidacy following Renny Ottolina's death. Arria, who had been Pérez's minister of information and tourism, failed to generate much support despite an impressive campaign chest and the hiring of yet another U.S. media expert, Patrick Caddell. Caddell utilized a middle-class-oriented Common Cause (*Causa Comun*) organization such as he used in 1973 to gain "independent" votes for Carlos Andrés Pérez. Charges of corruption and another tragic death, involving his chauffeur, doomed his campaign. Arria gained just short of 2 percent of the vote.

National Civic Crusade (CCN)

The fact that the National Civic Crusade (*Cruzada Civica Nacionalista*) was the 1968 electoral arm of the Pérez Jiménez movement made it more than an electoral protest movement. Had the CCN proven to be a major political force, gaining 20 percent to 25 percent of the national vote in 1968, the result could easily have been the return to power of the former dictator or the bloody repression of extraelectoral efforts to return him to power. The fact that the CCN's 1968 vote was only 11 percent of the total and that the various riots the CCN organized in 1969 to demand the seating of the former dictator as a national senator were relatively small and easily contained by the police ended the possibility of an early return to caudillo rule. The CCN legislative ballot vote in Caracas was an impressive 27 percent of the vote. The former dictator was elected to the Senate, only to be denied his seat due to his conviction for illicit enrichment.

The political arm of the Pérez Jiménez movement had, however, taken on an identity distinct from that of those engaged in plots and fantasies of military coup. The CCN leaders elected on the legislative ballot chose to take their seats in the national congress, state assemblies, and city councils in spite of the fact that the former dictator was denied his senatorial seat.[48] What has been labeled the Perónist wing of the CCN, led by deputy Abdelkader Márquez, broke from the essentially middle-class-oriented leadership of the party in 1970 to form the *Movimiento Popular Justicialista*. In reaction to this split in the CCN, Pérez Jiménez pressured from his exile in Madrid to have the two most distinguished CCN leaders, Erwin Burguera and Luis Damiani, removed from their party positions.

Many observers believe that the large 1968 CCN vote was another version of the traditional urban antiparty vote, not necessarily a vote for a return to dictatorship. Others believe that Pérez Jiménez wanted to create a personal cult focused on himself in his self-imposed exile and had no further need for a legal party because his appeal had been basically an antiparty one.

Thee was some talk in the early days of the 1973 precampaign that a clash of titans, Rómulo Betancourt vs. Marcos Pérez Jiménez, would take place. For a variety of reasons, Pérez Jiménez opted not to be a candidate. No less than three Pérez Jiménista candidates appeared. In 1973, the CCN candidate and other "independents" such as Pedro Tinoco, Jr., who professed some sort of relationship to the former dictator, had a minimal electoral impact. The CCN legislative vote in 1973 was 4.3 percent. Pedro Tinoco's own *Desarrollista* (Developmentalist) candidacy earned 29,339 votes in 1943. Pablo Salas Castillo, a founder of the CCN, tried to get the former dictator's wife to run for president. In desperation, Pablo Salas Castillo ran for president in 1978, receiving very few votes. A second, would-be surrogate for Pérez Jiménez created the United Nationalist Front in 1978. A third Pérez Jiménista candidate in 1973 was Gen. Garcia Villasmil. It appears that while the former dictator's electoral support has faded, he continues to do reasonably well in public opinion polls.

Military Candidates

Starting with the effort of retired Gen. Garcia Villasmil, a former defense minister, to call for a strong government free of party involvement in 1973, a number of other retired armed forces officers have toyed with the idea of launching a military-image, antiparty, electoral effort. Garcia Villasmil's 1973 vote on a Social Democratic Party label was minimal.

Two, apparently well-funded and well-organized, military candidates appeared on the scene for the 1983 elections. Both, it should be pointed out, were retired. Unlike the Admiral Larrazabal candidacy in 1958 and even that of Garcia Villasmil in 1973, the National Association of Retired Senior Military Officers (ASOGENAL) has played a factor in these two candidacies.

The post-petroleum bonanza in Venezuela in which crime, prices, unemployment, and government corruption have all risen, has put the public in a rather ugly mood regarding the political parties. The Gallup Poll in Venezuela in March and December 1980 reported 25.4 percent and 28.1 percent respectively preferring a "strong military government" (organized similar to those existing in other Latin American countries) to the present "democracy such as ours" ("with its crises and disorders").[49] A June 1981 poll prepared by the polling company *Mercanálisis* found 14 percent interviewed preferring a "military candidate," 10 percent a military dictatorship vs. 26 percent for an AD candidate, 22 percent for a COPEI candidate, and 12 percent for a Socialist one.[50]

The failed diplomatic effort to resolve the maritime-limit dispute with Colombia, the growing Venezuelan concern with the geopolitics of the Caribbean, and its territorial demands against Guyana have all placed national security and defense issues at center stage. Fear of crime on the streets, renewed guerrilla activity in the countryside, and the quiet Colombian invasion have also increased the concern for a strong government.

Two retired generals have answered the call. Former Defense Minister Luis Enrique Rangel Borgoin, a respected military intellectual and president of the Venezuelan Institute of Geopolitical Studies, has organized a new party, National Rescue, to support his candidacy. Gen. Rangel Borgoin has long advocated a policy of decentralizing economic activities from Caracas to border areas as the best means of securing national territory. A U.S. scholar's analysis of contemporary geopolitical thinking in Venezuela termed his views "temperate" and "conciliatory" regarding frontier issues. According to Judith Ewell, "He also publicly opposed the constitutional change which would allow the armed forces to vote."[51] As a presidential candidate, Borgoin sounded quite chauvinistic with regard to pressing Venezuela's claims to two-thirds of Guyana.

Rangel Borgoin is also the brother of Domingo Alberto Rangel, founder of MIR in 1960 and long a radical gadfly in Venezuela's leftist politics. Alberto Rangel had sounded the alarm in Venezuela in the late 1970s to prepare for a possible Brazilian military move to sieze the Orinoco Heavy-Oil Belt and the resource-rich Guayana region.

The second retired-general-turned politician had a lot more people worried. He is retired Gen. Arnaldo Castro Hurtado. Castro Hurtado resigned in disgust in 1979 as army chief of staff in an apparent disagreement with President Hererra Campins over the granting of amnesty to guerrilla leader Julio Escalona.[52] Castro Hurtado was said to have expressed the anger and frustration of many standing army officers who witnessed the triumphant return to public life of the hated radicals who had been responsible for the death of some of their fellow officers.

Retired Gen. Castro Hurtado was quite explicit concerning his criticism of the two parties of the populist establishment: AD and COPEI. His May 1979 resignation and the long line of military officers who showed up at his home at that time caused considerable speculation and rumors that a military coup was a possibility. According to some sources, the personal intervention of former Presidents Rómulo Betancourt, Carlos Andrés Pérez, and Rafael Caldera was a factor in calming down the situation.[53] Retired Gen. Castro Hurtado publicly insisted that several of the legal leftist parties function primarily as political covers for violent guerrilla activities. In addition to the Socialist League, he also had harsh words to say about MEP's Salom Meza Espinosa.

In June 1981, a new antiparty, antipopulist group of young executives and professionals who created the New Generation Movement nominated Castro Hurtado as their presidential candidate. The Castro Hurtado–New Generation pact ended with the news that Hurtado also contacted Douglas Bravo. The New Generation had gained the reputation of a rightist, technocratic-oriented movement. For some, the joining of forces between the retired officers of ASOGENAL and the civilian managers of the New Generation Movement constitutes a major challenge to the survival of the present populist electoral system. A young businessman, José Mariscobetre, appears to have emerged as New Generation's president and potential presidential candidate.

One Venezuelan official in a private conversation with the author called the New Generation Movement "fascist." The New Generation, on its part, has concentrated on winning adherents from the more conservatively inclined members of COPEI. Douglas Bravo's mid-1981 hint of a possible endorsement of Castro Hurtado illustrates the complexity of alliance building in Venezuela.

Because of Castro Hurtado's strong antipathy toward the Cuban government of Fidel Castro and his outspoken concern that many amnestied and active leftist politicians have retained one foot in the guerrilla tactic, some have used the pejorative label "fascist." That Castro Hurtado has associated himself with conservative rightists in the New Generation Movement is beyond question. One would be cautioned, however, not to quickly assume that this means that he is a fascist. As a young officer, he was characterized as pro-AD and was imprisoned briefly following the 1948 military takeover. He lived in exile during the Pérez Jiménez dictatorship.

According to one source, "He was a soldier of confidence of democracy in the struggle [in the 1960s] first against the dictatorial reaction and later the Castro communist insurrection."[54] He was in charge of the antiguerrilla operations and later director of Military Intelligence before becoming an army commander. This may account for his strong dislike of Castro and the retired guerrilla leaders. *Zeta*, a magazine with close ties to AD, called him a "tough soldier," while *Doble 6*, a magazine with ties to COPEI, called him "folkloric."[55]

The New Generation leadership insists that it is seeking a "new national convergence toward the center" and that it is not rightist. In 1980–81, its leaders—Vladimir Gessen (party secretary general), Maria Mercedes Pantin, José Marsicobetre (party president), and Egildo Lujan—could best be described as young middle-class managers seeking "a socially responsive, capitalism" rather than the strange mixture of capitalism and socialism that constitutes the reality of AD and COPEI. They have been especially hard on President Hererra, calling his version of populism "folkloric" and uniquely inept and inert.[56] New Generation leader, Mariscobetre's public statements in mid-1983 appealed to young Venezuelans to "use their intelligence."

The designation of Castro Hurtado as a "folkloric fascist" could be interpreted as one COPEI leader's polemical response to the New Generation challenge rather than to a real fear of fascism. It was expected that New Generation would damage COPEI more than AD in 1983, while MAS would damage AD more than COPEI. AD's response to the new image of Teodoro Petkoff and to his new respectability has been only slightly less heated than COPEI's attacks on the New Generation. AD's leaders take special umbrage at the comparison of Petkoff with Betancourt.[57]

The complex and contradictory nature of Venezuelan politics is illustrated by the fact that CCN elements, considered loyal to the former dictator, opted to align themselves with AD in the early 1970s and to support the AD candidate in 1978.[58] For those who believe in neat divisions between left and right, between dictator and democrat, and between good and bad, this news would be most disconcerting. Venezuela's political reality does, however, defy neat classifications.

NOTES

1. Howard R. Penniman, ed. *Venezuela at the Polls: The National Elections of 1978* (Washington, D.C.: American Enterprise Institute, 1980), p. 201.

2. The Gallup Poll of Venezuela in its December 1981 report revealed that AD's Jaime Lusinchi would have gotten 15.7 percent of the vote to Rafael Caldera's 11.6 percent. Regarding the legislative party ballot, the party preference was AD 29.3 percent, COPEI 15.3 percent, and 13.5 percent; "La Gallup de Diciembre: Caldera Pierde con Lusinchi y AD Aumenta su Ventaja." *El Nacional*, December 17, 1981, p. D-1.

3. "Rómulo Betancourt," *SIC*, Revista de Orientacion, Año VL (September-October 1981), 338.

4. Steven Ellner, "Comment: Leonardo Ruiz Pineda, Acción Democrática's Guerrillero for Liberty," *Journal of Interamerican Studies and World Affairs* 22 (August 1980), 390.

5. David E. Blank, *Politics in Venezuela* (Boston: Little, Brown, 1973), pp. 142–43.

6. One sympathetic account of these events is J.R. Nunez Tenorio, *La Izquierda y la Lucha por el Poder en Venezuela* (Caracas: Ateneo de Caracas, 1979). A much less friendly account is that of Benedict Cross (pen name) "Marxism in Venezuela," *Problems of Communism* 6 (November-December 1973). Steven Ellner's account relates the political-party dynamics to the guerrilla outbreak. See his "Political Party Dynamics in Venezuela and the Outbreak of Guerrilla Warfare," *Inter-American Economic Affairs* 34 (Autumn 1980), pp. 3–24. Robert J. Alexander also presents an account of these events in his biography, *Rómulo Betancourt and the Transformation of Venezuela* (Rutgers: Transaction Books, 1982).

7. Blank, *Politics in Venezuela*, pp. 174–76.

8. As part of Pérez's mania for good public relations, the Venezuelan presidency subsidized a number of foreign journalists and scholars in their "objective" endeavors to write about the wonders of Carlos Andrés Pérez. The late U.S. President John Kennedy's Press Secretary, Pierre Salinger, who had first met Carlos Andrés Pérez in 1961 while the latter was President Betancourt's interior minister, was the most notable journalist to accept Venezuelan government subvention, Pierre Salinger's *Cuadernos Venezolanos: Viajes y Conversaciones con Carlos Andrés Pérez* (Barcelona, Spain: Seix Barral, 1978). Salinger's book has been severely criticized in Venezuela. It is hoped his reporting for ABC News has a stronger relationship to reality. The production of critical studies of Carlos Andrés Pérez has assumed major importance. See Luis Lander, "Prologo," in *CAP 5 Anos: Un Juicio Critico* (Caracas: Ateneo de Caracas, 1978), Domingo Alberto Rangel, *Elecciones 1973: El Gran Negocio* (Valencia: Veni Vadell Hermanos, 1974).

9. John Martz, "The Evolution of Democratic Politics in Venezuela, in *Venezuela at the Polls: The National Elections of 1978*, ed. Howard Penniman (Washington, D.C.: American Enterprise Institute, 1980), pp. 23–24.

10. David E. Blank, "Oil and Democracy in Venezuela," *Current History* 78 (February 1980).

11. Enrique A. Baloyra and John D. Martz, *Political Attitudes in Venezuela: Societal Cleavages and Political Opinion* (Austin: University of Texas Press, 1979), p. 51.

12. Ibid.

13. See the study edited by Howard R. Penniman, *Venezuela at the Polls: The National Elections of 1978* (Washington, D.C.: American Enterprise Institute, 1980).

14. Juan Pablo Pérez Alfonzo outlived his philosophy regarding socialism, Third World development, technologically dependent oil, and the "New International Economic Order" in a book he wrote with a young Venezuelan scholar. See Juan Pablo Pérez Alfonzo and Ivan Loscher, *Alternativas* (Caracas: Garbizu and Todtmann, 1976). A polemical interview–dialogue between Juan Pablo Pérez Alfonzo and Domingo Alberto Rangel was published as a history of post-1958 Venezuela enitled, *El Desastre* (Valencia: Vadell Hermanos, 1976). For an account in English of Pérez Alfonzo's role in the creation of OPEC, see Kim Fuad, "Venezuela's Role in OPEC: Past, Present and Future," in *Contemporary Venezuela and Its Role in International Affairs*, ed. Robert D. Bond (New York: New York University Press, 1977).

15. Alfredo Tarre Murzi, "El Carlismo," guest editorial, *El Nacional*, July 12, 1980.

16. Donald L. Herman, *Christian Democracy in Venezuela* (Chapel Hill: University of North Carolina Press, 1980) pp. 92–95.

17. See Robert E. O'Connor, "The Electorate," in *Venezuela at the Polls: The National Elections of 1978*, ed. Howard R. Penniman (Washington, D.C.: American Enterprise Institute, 1980), pp. 56–90, for an analysis of voter perception of the 1978 issues.

18. Blank, "Oil and Democracy in Venezuela."

19. Daniel H. Levine, "Democracy and the Church," in *Venezuela: The Democratic Experience*, ed. John D. Martz and David J. Myers (New York: Praeger, 1977), p. 167.

20. Ibid., p. 108.

21. Alfredo Pena, "Foro: Valmore Acevedo Amaya," *El Nacional*, February 18, 1980, p. 4.

22. Charles D. Ameringer, "The Foreign Policy of Venezuelan Democracy," in *Venezuela: The Democratic Experience*, ed. John D. Martz and David J. Myers (New York: Praeger, 1977), p. 339.

23. John Gerass, *The Coming of the New International: A Revolutionary Anthology* (New York: World, 1971), p. 423.

24. Cross, "Marxism in Venezuela," p. 64.

25. "Sign of Venezuela's Maturity: Guerrilla in Congress," New York *Times*, November 25, 1980, p. A2, and Steven Ellner, "Venezuelan Surprise, That of Friendly Marxists," New York *Times*, August 25, 1981, p. 19.

26. One such charge by AD leaders appeared in "El MAS y COPEI Rebajaron Medio Millon a Empresa de Luben Petkoff," *El Universal*, December 18, 1981.

27. Luciano Pellicani, *Gramsci, An Alternative Communism?* (Stanford, Calif.: Hoover Institution Press, Publication 243, 1981), p. 8.

28. See the Friedrich Ebert Foundation publication, *Horst Emke, Democratic Socialism and Eurocommunism: The Policy of Detente and Ideological Controversy* (Bonn-Bud Godesberg: Federal Republic of Germany, 1977), p. 18.

29. John Martz, "The Minor Parties," in *Venezuela at the Polls: The National Elections of 1978*, ed. Howard R. Penniman (Washington, D.C.: American Enterprise Institute, 1980), p. 158.

30. David J. Myers, "Venezuela's MAS," *Problems of Communism* (September-October 1980), 27.

31. Rómulo Betancourt made these charges in a July 15, 1967, letter he sent to AD leaders from Naples, Italy, Alexander, *Rómulo Betancourt*, pp. 601–03.

32. Franklin Tugwell, *The Politics of Oil in Venezuela* (Stanford, Calif.: Stanford University Press 1975), pp. 21–121.

33. Ibid., p. 106.

34. "Se Mueven ex Militares de Venezuela Hacia la Formacion de un Partido Politico," *Diario de Las Americas* (Miami), May 29, 1981.

35. *El Diario de Caracas*, guest editorial, José Vicente Rangel, "La Candidatura Militar, un Test Paro Politicos," April 7, 1981.

36. *Resúmen* magazine has periodically used the Gallup Organization to poll the competitive fortunes of Petkoff and Rangel. Its issue of December 20, 1981, indicated that Petkoff's support was "dynamically expanding."

37. "Hector Mujica un Communista que no esta Fuera de Epoca," *El Diario de Caracas*, February 8, 1981.

38. Interview with Pedro Ortega Díaz, PCV Political Bureau, *World Marxist Review* (March 1981), 32.

39. "China: A Zigzag in Search of Socialism," *Decada 80* (May 1980).

40. Myers, "Venezuela's MAS," p. 24.

41. David E. Blank, "Venezuela," in *1981 Yearbook On International Communist Affairs*, ed. Richard F. Staar (Stanford: Hoover Institution Press, Publication 250, 1981), p. 117.

42. Centro Gumilla, *Los Partidos Políticos de Venezuela*, 2nd ed. (Caracas: Centro Gumilla, Curso de Formacíon Socio Político 2, 1977), p. 31.

43. Gerass, *The Coming of the New International*, p. 425.

44. Militar Retirado Venezolano Preocupa a los Jefes Políticos, *Diario de las Americas*, July 7, 1981.

45. Frank Bonilla and Philip Raup, Jr., "Social Background Data in Elite Analysis: A Venezuelan Example," (Paper presented at the annual meeting of the American Political Science Association, Chicago, September 5–9, 1967), p. 8.

46. Jesús Angel Paz Galarraga, "Caracas, Una Incognita Electoral," *El Nacional*, April 20, 1967, p. 4.

47. Martz, "The Minor Parties," p. 164.

48. The history of the post-1958 Pérez Jiménista movement has largely escaped U.S. scholarship. One notable exception is Judith Ewell, "Politics and Pérez Jiménistas in the 1960's and the 1970's" in *Indictment of a Dictator: The Extradition and Trial of Marcos Pérez Jiménez*, ed. Judy Ewell (College Station: Texas A. & M. University Press, 1981), Chapter 8.

49. Instituto Gallup de Venezuela, *Indice Gallup de Opinion Publica* 4 (December 1980), p. 79.

50. Gene Bigler, "Political Restraint and Military Professionalization in Venezuela," (Paper presented at the conference on the New Military Politics in Latin America, Hoover Institution, Stanford, Calif., August 1981), p. 47.

51. Judith Ewell, "The Development of Geopolitical Thinking in Twentieth Century Venezuela," (Paper presented at the 1980 Latin American Studies Association meeting, October 16–18, 1980, Bloomington, Ind.). Ewell calls Bourgoin "more temperate than Lt. Col. López Sanchez (p. 16); "General Bourgoin took a conciliatory line on frontiers and geopolitics; he also publicly opposed the constitutional change that would allow the armed forces to vote" (p. 31).

52. David J. Myers, "The Elections and the Evolution of Venezuela's Party System," in *Venezuela at the Polls: The National Elections of 1978*, ed. Howard R. Penniman (Washington, D.C.: American Enterprise Institute, 1980), pp. 225–26.

53. Ibid., p. 226.

54. "Castro Hurtado en Posicion Estratgica," *Zeta*, December 6, 1981.

55. Ibid., and *Doble 6* 3 (December 4–10, 1981), p. 16.

56. See the New Generation advertisement, "El Gobierno de la Oposicion y la Oposicion del Gobierno," *El Diario de Caracas* 1 (August 10, 1980).

57. Valetin Iglesias, "Pico y Pala: Teodoro Petkoff," *El Nacional*, July 8, 1980, p. A5.

58. Ewell, "Politics and Pérez Jiménistas," p. 157, 164.

5

POLITICAL VIOLENCE: THE PRAETORIAN IMPULSE

The continuing attraction in Venezuela of the myth of a simple, violent solution to complex problems has maintained the praetorian option with two rather different faces. For some, there is the belief that a "benevolent authoritarian" exists among the senior officers of the armed forces. Brazil since 1964, especially its impressive economic performance in the 1967–75 period, is their model. For others, there is the belief that a "benevolent" and heroic leader can emerge from a revolutionary process to do what needs to be done. Cuba's Fidel Castro and the myth of the late Ernesto "Che" Guevara is their model. The first praetorian submyth is that of the classical coup while the second is that of the revolutionary insurgency. In a very real sense, the attractiveness of the praetorian revolutionary impulse is not political; it is based on a belief in magic.

Before 1936, the principal means utilized by contenders for national power in Venezuela was the aggregation of armed force sufficient to defeat the force organized by actual or potential competitors. Since 1936, political violence has been just one of the strategies open to Venezuela's various power contenders. In 1959, Rafael Caldera, who served as president in the 1969–74 period, said "Venezuelans are so accustomed to make the army the arbiter of their political contests, that at each moment the most varied groups for the most dissimilar ends attempt to involve the army in new adventures to change our political reality."[1]

In 1945, the AD leadership did not hesitate to utilize military conspirators in its drive for power. Three years later, significant civilian elements approved the military overthrow of the mass-elected AD government. Finally, the 1958 overthrow of the Pérez Jiménez dictatorship was the result of divisions within the military as well as maneuvers by civilian groups. Historically, revolutionary violence has been a normal means of politics in Venezuela.

Between 1958 and 1964, attempts were made both by advocates of radical revolution and by advocates of conservative reaction to replace the fledgling representative government with one based upon force. Since 1964, violence has played a more limited and contained political representational role.

A variety of political elements continue to use force and violence to demonstrate the seriousness of their demands. Not all displays of force are intended to be violent. Often they are meant only to demonstrate the credibility of a group's claim that it should be considered among the important power contenders. In October 1979, AD politicians organized a boisterous and, some say violent, labor demonstration in Caracas as a means of demonstrating that the party retained its control over the labor movement.

In post-1958 Venezuela, the most significant incidence of violence occurred during the upheaval between the January 23, 1958, coup and the inauguration of President Raúl Leoni in early 1964. The forces opposed to the return to elected civilian government sought a military coup. Those wanting a radical transformation of Venezuela fell under the sway of the myth of the Cuban revolution.

SUBSYSTEM ONE: CLASSICAL COUP

The military and its ability to coercively apply the brakes to politics was utilized in part by antipopulist elements in the early 1960s period of upheaval in a failed effort to terminate the post-1958 democratic experiment. For a small but strategically placed coalition of military officers and civilians, there was an apparent pressing need to replace the open and low-coercion political style of the emergent system with one that was more closed and repressive. During these failed attempts at classical military coups, the vast majority of military commanders remained loyal to the pluralistic and open political system.

The military's present role is that of a praetorian subsystem that has been well-rewarded for its defense of the ongoing system. It has been obedient to the rule of law even when that has embarrassed its commanders in their simulated war games with Guyana.

Classic Coup Attempts

Civilian opposition to the early return to representative government and the legalization of AD in 1958 focused on the military men loyal to the new minister of defense, Col. Jesús Maria Castro León. Following the riots that greeted the visit of Vice-president Richard Nixon and the resultant government crisis, Castro León led an attempted coup against the provisional government on July 22, 1958. One of the principal causes for its failure was the mobilization of mass demonstrations and strikes in favor of restoring representative democracy. The same popular opposition was aroused on September 7, 1958, when another conservative military officer attempted to seize power and to prevent the elections scheduled for December of that year.

Following the inauguration of President Betancourt, conservative military factions seriously threatened to overthrow his government in October 1958, January 1960, April 1960, and June 1961. The April 1960 coup attempt was led by Castro León who, with the help of the late dictator of the Dominican Republic, Rafael Leonidas Trujillo, had returned from exile. Finally, in June 1960, President Betancourt narrowly escaped assassination in a plot that apparently had the backing of Trujillo.

The form of political violence undertaken by rightist military and civilian opponents of both provisional President Larrazábal and elected President Betancourt between 1958 and 1961 was the classic coup. The classic coup depends for its success upon the quick seizure of power and the presentation of a *fait accompli* to an essentially uncommitted and largely neutralized populace and military.[2] The political environment during the years 1958 to 1960 was not propitious to a traditional application of political violence. All political factions, from the Social Christian Party (COPEI) to the Venezuelan Communist Party (PCV), united in insisting that representative democracy be restored.

One big reason for the failure of these initial coups and the few subsequent rumored efforts since 1965 has been that key business associations such as the Federation of Chambers of Industry and Commerce, the Roman Catholic hierarchy, and the military high command all opposed them.

The Military's Present Role

Barring a major economic breakdown and precipitous rise in social tension and violence in Venezuela, it is unlikely that the military command will radically lessen its coexistence with populist democracy. For some economists, Venezuela has in fact entered a period of crisis with the decline in the world price of oil in 1981.

Whatever Venezuela's long-term economic future, the next few years should bring double-digit inflation, high unemployment, and a growing sense that public services are deteriorating. There is also a growing awareness of geopolitics and national security. The Falkland Islands tragedy of 1982 has focused attention on Venezuela's claim to two-thirds of the territory of its neighbor, Guyana. The tension with Colombia increased sufficiently in 1980 that public-opinion polls revealed that this neighbor democracy and fellow Andean Pact member had replaced Cuba as Venezuela's perceived worst enemy. There was also a revival of the terrorist hijacking of planes and the outbreak of guerrilla activities in 1980, 1981, and 1982.

In this climate, the military as an institution has notably increased its visibility and influence within the total Venezuelan political picture. The *VI National Plan 1981–85* was the first such four-year planning document to contain a chapter on national security and defense. The National Security and Defense Council, created by law in 1976, has launched an ambitious program of public education on the subject.[3] Venezuela's military has also become a factor in the nation's industrial economy through the establishment of a number of military-run state enterprises that manufacture and maintain weapon systems. "Finally, the general consciousness raising about national security and defense led to the founding of a national war college [Institute for Advanced Studies in National Defense, IAEDN]."[4]

There are some in Venezuela who fear that the activities of IAEDN, in developing a unified geopolitical and national-development consciousness among military officers and business managers, could lessen the tolerance of these groups toward the muddling-through mentality of democratic populism. An IAEDN war game caused a major constitutional crisis in Venezuela in 1981 when a reporter who wrote an article about it was arrested for violating national security. The Supreme Court in a December 1981 decision, however, clearly established the supremacy of the civilian judicial system over that of the military and the sanctity of freedom of the press.[5] This decision was acclaimed by all civilian sectors and accepted by the military.

While Venezuela's record on human rights since 1964 has been exemplary, the use of military tribunals and military arrests of civilians has caused much concern. The Organic Law of Security and Defense allows the president to declare security zones in which the armed forces can impose "exceptional means" of preserving order. Charges have been made that this law has been abused in order to break labor strikes against key state-owned enterprises. This has been the case in the planned industrial city of Ciudad Guayana and in the nearby workers camp that houses those building the massive Raul Leoní-Guri Dam.[6]

The administration of Carlos Andrés Pérez was accused by its opponents of flagrantly abusing the use of military justice. People's Electoral Movement (MEP) Deputy Salom Mesa, a former AD colleague of President Pérez, was subjected to military justice because of alleged involvement in the still-confused story of the kidnapping of U.S. businessman William Niehouse. His arrest took place despite his congressional immunity. Salom Mesa was reelected in 1978 and released from military imprisonment. The COPEI administration of Hererra Campins (1979–84) has been identified with the retreat of military justice from the civilian sector.

The military high command was embarrassed in late 1981 by the disclosure by a colonel who had been passed up for promotion that he may have been blackballed because he had previously tried to denounce irregularities in military purchases while in command of logistics for the strategic Gran Sabana region facing Brazil and Guyana. The presidential candidacy of former Defense Minister Rangel Bourgoin may be adversely affected by this revelation since he had been the Defense minister when the original charges were made. The military's reputation was also to be stained by the persistent rumors of impropriety in lobbying for the purchase of 18 advanced U.S. F-16 fighters and some British aircraft.

The praetorian subsystem of Venezuela's politics has increased in importance. It has done so, to a degree, at the expense of the representative-democratic subsector. Popular disillusionment with democracy is also on the increase. The tripartite model of Venezuela's politics presented in Chapter 1 suggests an elasticity and flexibility in the relationship among the three political subsystems. When faced with a crisis of a national-security nature, the electoral subsystem may decline in influence while that of praetorianism may increase. However, once this crisis has passed, the electoral subsystem in Venezuela has been able to reassert its centrality. Only a major, violent, political breakdown could convince Venezuela's military leaders to end their restraint.

SUBSYSTEM TWO: REVOLUTIONARY INSURGENCY

In late 1960, a most serious revolutionary threat to the ongoing populism emerged. Youthful elements on the political left attempted to emulate the Cuban experience in which a few hundred guerrillas led by Fidel Castro caused the collapse of the Fulgencio Batista dictatorship. These youth, who were first attracted to violence in the 1959–60 period, were not members of the PCV but of the two reform-oriented populist parties: the Democratic Action (AD) and Democratic Republican Union (UD). They had directly experienced participation in the revolutionary ferment of the early 1958 overthrow of the Marcos Pérez Jiménez regime and the spontaneous rioting that greeted Vicepresident Richard M. Nixon in May of that year. For these youths, Venezuela was ready for a radical transformation by means of a mass-based revolutionary movement.

Venezuela's cities were filled with slums that housed unskilled, recent immigrants from the countryside who were filled with rising aspirations and who, in the period of late 1950s' oil glut, faced lowered job prospects. The universities were also filled with aspiring young leaders who saw the emerging Cuban model as a means of personal and national fulfillment. There was a natural youthful exaggeration of the role popular demonstrations had played in the downfall of Pérez Jiménez and in thinking that the spirit of January 1958 would become a permanent part of Venezuelan politics. For some, studying for one's engineering examinations seemed considerably less challenging than leading another demonstration. The fact that the campus of the Central University of Venezuela occupied the geographic center of Caracas so that student demonstrations would quickly divide the city and paralyze its movement also engendered a sense of heightened importance.

Fidel Castro took power in Cuba on January 1, 1959. The Cuban revolution succeeded between Rómulo Betancourt's election to the presidency of Venezuela in December 1958, and his inauguration in March 1959. Although much of their rhetoric was similar and Castro and Betancourt found grounds to cooperate against a common enemy, Rafael Leonidas Trujillo, in 1959, the Cuban and Venezuelan situations were quite dissimilar. The total collapse of Batista's army enabled Castro to consider the grand transformation of Cuban society.

Betancourt, on the other hand, recognized that his rural-based electoral victory could be annulled by Venezuela's still-intact armed forces if his government proved to be a disruptive one. In fact, Rómulo Betancourt was principally concerned with reconciling his party, AD, with the two other major political parties, COPEI and the URD, and with nonpartisan power contenders like the church, the armed forces, and the business community. The grand transformation of Venezuela via a program of revolutionary reforms was not contemplated by the AD leadership. AD leaders were not involved in the January 1958 overthrow of Pérez Jiménez, and therefore AD could not claim exclusive guardianship of the spirit of 1958 as Fidel Castro did of the struggle against Batista.

The origin of the break between those who would have emulated the Cuban experience and the tripartite coalition government led by Betancourt was the government's decision to make its reforms within a framework acceptable to the domestic business community and to international agencies like the International Monetary Fund and World Bank. The government considered the support of these financial agencies crucial to the success of its development plans. In addition, the support of these international agencies helped the government acquire an aura of respectability that was important in maintaining the loyalty of key military and business groups.

In early 1960, the government announced both the acceptance of loans from a New York consortium of private banks and from the International Monetary Fund. In April 1969, the radical youth of AD split from the principal government party to form their own Revolutionary Left Movement (MIR).

At the September 1960 meeting of the Organization of American States (OAS), the Betancourt government agreed to break diplomatic relations with Fidel Castro's Cuba in return for U.S. support of sanctions against the Dominican Republic's dictator, Trujillo. The move greatly increased tension between the government and its left-wing opponents.[7] In November 1960, a series of urban riots, the first bloody clash between the government and its pro-Castro opposition, took place; four died in these clashes. The URD, under pressure from its own left wing, abandoned the government.

The degree to which the leftist insurgents operated according to some grand design and the degree to which the Castro government was involved in the formulation of this design is in dispute. Former insurgents and supporters of the Betancourt government remain adamant in their insistence that the other side initiated the violent confrontation.

Observers sympathetic to the guerrillas insist that the Betancourt government initiated the violence by its attempted repression of its leftist opposition. According to this view, the radical left launched its own insurgency only in response to this repression. These sources cite the periodic suspension of constitutional guarantees to support their charge of government repression. An account of the period favorable to the government insists that these limited constitutional restrictions were necessitated by the extreme violence of the insurgents.[8] Those sympathetic to the Betancourt government insist that the radical youth of both the MIR and the PCV started the violence. According to the government, the radical youth tended to misread Castro's success in Cuba.

Since revolution had been a respectable political process in Venezuela, many young radicals thought it was their turn to have a try at it. The radical youth may also have despaired of bringing about the revolutionary transformation of Venezuela through electoral politics. The political parties with which they associated, MIR and PCV, did not seem to have the ability to become major contenders in the electoral arena.

A study done for the U.S. Defense Department divided the revolutionary insurrection of the early 1960s into two phases.[9] The first phase reportedly was aimed at a rapid victory, and the second received the label "protracted struggle." According to this study, the elements involved in bringing about a rapid victory for the revolutionaries included urban riots, military rebellions, and rural-based guerrilla activity. The second phase, the protracted struggle, was largely a rural-based guerrilla operation.

The urban insurgency began in late October 1960 with a series of riots and student and working-class strikes. It was in this initial phase that the maximum numbers participated. In 1961 alone, 113 distinct, significant urban incidents involving thousands of people occurred; nine people were killed and 39 reportedly wounded.[10]

Recruitment into this urban insurrection was unquestionably aided by the existence of large numbers of unemployed people in the slums, or *barrios*, during the period of austerity and budgetary retrenchment the Betancourt government initiated in 1961. Unemployment in the barrios of Caracas, Valencia, and Maracay was at least twice that of the national average of from 9 percent to 14 percent. In petroleum communities, such as the city of Cabimas, the slowdown in petroleum production apparently caused unemployment to soar to the phenomenal rate of 70 percent.

The nostalgic claim made in 1968 by Teodoro Petkoff, then a young Communist guerrilla leader, that in 1962 "the hills of Caracas became fortresses and were nightmares for police and troops where government searches yielded next to nothing . . . because the vanguard was closely linked with the people in the area," was not entirely an exaggeration.[11] The combination of strikes and urban demonstrations was apparently building to a crescendo when in May and June of 1962 two leftist-associated military rebellions occurred in Carupano in the eastern region and Puerto Cabello in the central region. The death toll reported in the Carupano mutiny was 40, and in the prolonged street fighting in Puerto Cabello, about 400. One PCV national deputy, Eloy Torres, was captured fleeing Carupano along with some of the leaders of the rebellion. MIR leader Simón Sáez Mérida was implicated by the government in the Puerto Cabello rebellion.[12] The failure of the two military mutinies in 1962 ended what popular support existed for the radical insurgency.

In 1961, a number of rural guerrilla groups, made up almost entirely of university and secondary-school students, began to operate. Only in two areas, the Falcón and Trujillo states, where the guerrilla leaders were local boys with family connections in the region, did guerrilla bands survive their initial defeats. Unlike the urban insurrection, rural guerrilla groups ran into a hostile peasantry that generally aided the military in hunting them down. The reports of numerous, scattered guerrilla clashes—even though they failed—added to the general feeling of alarm and the impression of government paralysis, and so they aided the insurgents.

The defeat of the military mutinies in Carupano and Puerto Cabello, along with the guerrillas' utter failure to establish a network of sustained operations, ended the insurgents' hope for a rapid victory. It is difficult to estimate the degree to which bad planning and bad luck determined this failure. Had the two military uprisings been better coordinated and simultaneous, perhaps other military units would have followed suit. The revolutionaries' failure to engender mass support for their cause and the adoption of terrorist tactics to attempt to sabotage the 1963 elections were major strategic blunders.

Scholarly analysis of the two 1962 military uprisings as well as published accounts written by the now-amnestied military conspirators indicate that the young marine officers who led the coups had personal career-advancement goals rather than revolutionary ones. In other cases, they may have considered themselves to be potential Venezuelan versions of Argentina's Juan Perón or Egypt's Gamal Nassar.[13]

After 1964, the revolutionary insurgency rapidly declined and degenerated into sporadic terrorism. The 1967 ambition of Cuba's Fidel Castro to project himself as a potential leader of the Third World and the break between the PCV and Cuba may have convinced many guerrilla leaders that they were being used as pawns in a geopolitical struggle that had little to do with enhancing equality and fraternity in Venezuela. In addition to reflecting the international ambitions of Fidel Castro, the issue of armed struggle became involved in the Soviet-Chinese dispute over hegemony of the World Communist circle of nations. The return to open politics by means of accepting amnesty (and government largess) was underway.

The Soviet Union in the late 1970s was interested primarily in maintaining its hegemony over the World Communist movement and in détente. It was not interested in having a number of freewheeling revolutionary leaders appear on the Third World scene to challenge its ideological hegemony and its investment in détente. For the Soviets, the urban working class and the pro-Moscow Communist parties were the vanguard of the revolution. For the Maoists, then in control of the People's Republic of China, the rural peasantry and the pro-Chinese Communist parties were the vanguard of the revolution.

Cuba's Fidel Castro introduced his own vanguard thesis, by his sponsorship of a book written by Frenchman Régis Debray in 1967, that there is a third way to make a revolution.[14] For Debray, Castro, and Guevara, the job of a revolutionary was to make a revolution and not to wait for either the urban worker or rural peasant. Debray and his Cuban sponsors glorified violence and made the professional, isolated guerrilla the exclusive vanguard of the revolution. Rather than having an armed struggle only when needed to defeat repression, the armed struggle was to be the means of radicalizing the population rather than mass mobilization via labor unions, street demonstrations, or peasant invasions of land.

The Debray-Castro-Guevara thesis of the late 1960s relied on the alienated youth (rather than on organized workers and peasants) to provide the primary manpower for the revolutionary struggle. It also hoped for a union of revolutionary and nationalist fervor. It has, however, been severely criticized as falling into the traditional Latin-American caudillo mold, for being exclusively elitist, and for being outright fascist.[15] The myth of revolutionary violence may serve these youth as a welcomed therapy for their redundance and perceived disinheritance in a time of rapid change; however, similar to the drug scene, it has proven to be a very self-destructive therapy. For the "déclassé intellectual" leader, the myth and the utilization of it may be used as an instrument to break into the system.

In mid-1980, outbreaks of violent guerrilla and terrorist activities, caused primarily by a group calling itself Red Flag (*Bandera Roja*, BR), began. The arrest in 1980 of Commander "Zerpa" (Pedro Véliz) made Gabriel Puerta Aponte supreme commander of the Bandera Roja. Gabriel Puerta Aponte, in his late 30s in 1981, apparently began his political career as an AD student leader who joined the 1960 MIR defection. The origin of the BR was the continuing guerrilla operations in the eastern states of Venezuela called the Antonio José de Sucre Guerrilla Front (FGAJS). In 1977, FGAJS split as two of its leaders, Julio Escalona and Marcos Gómez, began their return to open party politics and the creation of the *Liga Socialista* (LS). Those who remained committed to the armed struggle renamed their group Bandera Roja. Gabriel Puerta Aponte became supreme commander as some would-be competitors died in action, accepted amnesty, or decided to form their own group.

In January 1981, Puerta Aponte promised a revolutionary offensive. In December 1981, BR apparently fulfilled its promise in a series of spectacular actions. On December 7, within the time frame of one-half hour, three domestic Venezuelan airlines were hijacked.

The planes eventually made their way to Cuba in order to facilitate the peaceful and safe return of the crew, passengers, and aircraft. The hijackers were arrested by the Cuban authorities, and Venezuela, despite the chill in its relationship with Cuba, indicated at that time that Cuba had acted correctly in this instance. Although no definitive accusation has been made against BR for this triple hijacking, informal leaks in 1982 placed the blame on this band. However, in late 1982, a top Venezuelan security official, Roberto Uzcatequi, indicated that the hijacking was the work of a "Caribbean tropical power."

While this triple hijacking received considerable international attention, the land operations of BR in late November and December 1981, according to some sources, heralded a new, dangerous, guerrilla insurrection that aims at disrupting Venezuela's development of the Orinoco Heavy Oil Belt. In late November and December 1981, a band of about 25 guerrillas attacked National Guard posts in three small communities near the Orinoco Oil Belt. One young national guardsman was killed.

Two basic theses have been presented to explain this guerrilla outbreak. As suggested, Puerta Aponte is seen as a pawn in a larger geopolitical struggle. According to adherents of this thesis such as Leonardo Montiel Ortega, an ally of AD, the BR attacks near the Orinoco Heavy-Oil Belt, called *la faja*, meant that the Soviet Union or Cuba was testing the feasibility of disrupting the development of the Orinoco Heavy-Oil Belt and destabilizing the Caribbean Basin as a means of denying the United States a secure hemispheric oil supply.[16]

A second variant of this thesis that perceives the BR to be part of a broad geopolitical strategy, suggested by retired general and 1983 presidential candidate Castro Hurtado, argues that the legalized, leftist former guerrillas and the existing guerrillas are in fact two strategic legs of the same international force. Given the reported claim by Puerta Aponte that he considers himself to be a Venezuelan version of the Yugoslavian Tito and Albanian Hoxa and given the evidence that the alleged legal arm of this endeavor, the LS, and Carlos Betancourt, former BR leader displaced by Puerta Aponte, were involved in the Qadafi-funded forum held at the Caracas Hilton, one may see a Libyan rather than Soviet or Cuban geopolitical hand behind the BR.

The second thesis suggests that the three late 1981 guerrilla attacks and the triple hijacking reflected the machismo phantasy of Puerta Aponte and his ego insistence to disrupt the political left from achieving a unified slate and presidential candidate for 1983. In this sense, the violence in 1981 merely reflected the internal caudillo struggle among diverse elements of the left. It had no greater geopolitical significance. According to LS leader and former BR colleague Marcos Gómez, the actions of Puerto Aponte "lamentably prejudice the process of [political] recuperation of the revolutionary forces . . . and aid the ultra right offensive, such as that of Castro Hurtado, which aspires to create a more authoritarian system."[17]

In the spring of 1982, Venezuelan counterinsurgency police captured Puerto Aponte. In a later clash, four BR guerrillas and two police officers were killed. In September 1982, Venezuela's most violent guerrilla battle since the 1960s took place near the Orinoco Heavy Oil Belt. At least 25 were killed in the clash. In September 1982, army and police attacks on the BR may have ended its operational capabilities, although some guerrillas evaded the military.

CONTINUING REPRESENTATIONAL VIOLENCE

Violence, force, or the threat of violence remain auxiliary aspects of the two other principal styles of politics: electoral politics and planning. The AD leadership in the 1968 electoral campaign attempted to destroy the People's Electoral Movement (MEP), organized by a breakaway faction of the party, by physically harassing local MEP leaders and rallies. In some cases, raids on MEP headquarters occurred, and Ministry of Public Works personnel and vehicles were frequently seen carting off MEP posters. In order to demonstrate that MEP had not been reduced to a puny and insignificant political element, its local leaders occasionally used force and counterharassment against their AD counterparts. These clashes never went beyond the fist-fight, street-brawl, and occasional-shot-fired-in-the-air stage. The MEP's ability to defend itself against such harassment was crucial to its remaining a major contender in the 1968 general elections.

Following the defeat of their own party, many AD leaders reacted strongly to the election of former dictator Marcos Pérez Jiménez as a senator from the federal district (Caracas). A number of AD leaders publicly declared that Pérez Jiménez would not be allowed to take his seat. This implicit threat of violence gained credibility in early January 1969 when a city councilman elected on the National Civic Crusade (CCN) ballot, Pérez Jiménez's party, was reportedly attacked by AD supporters on his way to be sworn in.

The Supreme Court of Venezuela later annulled the election of Marcos Pérez Jiménez on the grounds that the election law prohibits anyone who did not vote from being elected to a public office. Following his release from prison in August 1968 and the restoration of his political rights, Pérez Jiménez carried on his campaign from Spain and was not in the country on election day. In November 1969, CCN members forcibly disrupted a meeting of the national Senate to publicize their demand that Pérez Jiménez be seated. Many Venezuelans also feared that the Pérez Jiménistas hoped to create a climate propitious to a military coup.

The CCN supporters were more violent and disorderly in their November 1969 demonstration than the AD supporters had been in their January 1969 protest against the seating of Pérez Jiménez because the CCN had a lot less to lose by causing chaos. The AD, as a major establishment party, had to limit its violent demonstrations to fit the particular objective of denying a Senate seat to Pérez Jiménez. Had AD been more unrestrained in its protest, it would have risked being labeled a sore loser or even being charged with trying to nullify the entire election. The favorable response to all key political forces to AD's restrained action made it possible to legally deny the seat to Pérez Jiménez.

The 1981 violent actions of the BR may be seen as just one further example of a Venezuelan political group using violence as a means of representation and of demonstrating to others that one's concerns must be considered. One need not accept any geopolitical involvement in or threat to Venezuela to explain the continuing viability of violence. Both Montiel Ortega and Castro Hurtado have been accused by their opponents of exaggerating the significance of the BR in order to advance their own political aims.

There is an additional dimension to praetorianism that eludes systematic study, but which most observers of politics know exists. This is the linkage between the overt activities of the world of politics and business and the subterranean world of criminal activity. There is mention of the "Carmona case," "cocaine cowboys," and thugs being used to settle business and political issues. Venezuela is not unique in its possession of a politically astute underworld.

NOTES

1. Quoted in John J. Johnson, "The Soldier as Citizen and Bureaucrat," in *Latin American Politics: Twenty-four Studies of the Contemporary Scene*, ed. Robert G. Tomaset (New York: Anchor Books, 1966), p. 192.

2. This description of the classic coup comes from Edward W. Gude, "Political Violence in Venezuela, 1958–1964," (Paper presented at the annual convention of the American Political Science Association, Chicago, September 5-9, 1967), p. 13.

3. Gene E. Bigler, "Political Restraint and Military In Venezuela," (Paper presented at the conference on the New Military Politics in Latin America, Hoover Institution, Stanford, Calif., August 27–29, 1981), p. 43.

4. Ibid., p. 26.

5. This has been called the "case of Maria Eugenia Diáz," reporter for the progressive business newspaper, *El Diario de Caracas*. *El Diario de Caracas*, December 26, 1981.

6. Charles Anderson in his comparison of Betancourt's democratic reform strategy and Castro's revolutionary approach has suggested that perhaps the "presence of a revolutionary situation" in Cuba and its absence in Venezuela, more than the personal plans of these two leaders, determined their choice of strategies. See Charles Anderson, "The Strategies of Development Policy," in *Politics and Economic Change in Latin America: The Governing of Restless Nations*, ed. Charles Anderson (Princeton, N.J.: Van Nostrand, 1967). There were a number of AD-sponsored plots against the Pérez Jiménez regime, all notable failures. For a description of these attempts, see José Rivas y Rivas, *Historia Grafica de Venezuela, El Mundo y la Epoca de Pérez Jiménez*, vol. 2 (Caracas: Pensamiento Vivo, 1961).

7. Only after the United States voted to sever its diplomatic relations with the Dominican Republic and to place economic sanctions against it because of Trujillo's "aggression against Venezuela" (at the sixth meeting of OAS foreign ministers in San José, Costa Rica) did Betancourt agree to censure Cuba (at the seventh meeting of the OAS a few days later in the same city).

8. James Petras describes the insurgency from a viewpoint sympathetic to the guerrillas in "Revolution and Guerrilla Movements in Latin America: Venezuela, Guatemala, Colombia and Peru," in *Latin America: Reform or Revolution?*, ed. J. Petras and M. Zeitlin (Greenwich, Conn.: Fawcett, 1968). Robert J. Alexander describes the same events from a viewpoint sympathetic to the government in *The Venezuelan Democratic Revolution: A Profile of the Regime of Rómulo Betancourt* (New Brunswick, N.J.: Rutgers University Press, 1964), pp. 118–36.

9. Thomas L. Lalley, *Castro-Communist Insurgency in Venezuela* (Alexandria, Va.: Atlantic Research Corp., 1964).

10. Ibid., pp. 29–30.

11. Teodoro Petkoff, "Pre-election Climate in Venezuela," *World Marxist Review* 2 (April 1968). Philip B. Taylor in his study of the revolutionary insurgency in Venezuela conceded that the urban population did not cooperate with the government's counterinsurgency operations. See "Venezuela (1958–1963)," in D.M. Condit, Bert H. Cooper, et al., *Challenge and Response in Internal Conflict,* Vol. 3: *The Ex-*

perience in Africa and Latin America. (Washington, D.C.: American University Center for Research In Social Systems, 1968), p. 492.

12. Steven Ellner, "Political Party Dynamics in Venezuela and the Outbreak of Guerrilla Warfare," *Inter-American Economic Affairs 34*, Autumn, 1980.

13. See especially Ali Brett Martinez, *El Porteñazo: Historia de una rebelión*. (Caracas: Ediciónes Adaro, 1973) and Hugo Trejo, *La Revolución no ha Terminada* (Valencia: Vadell Hermanos, 1971).

14. Régis Debray, *Revolution in the Revolution?: Armed Struggle and Political Struggle in Latin America*. (New York: Grove Press Inc., 1967).

15. Lewis Coser, "Fanon and Debray: Theorists of the Third World," in *Beyond the new left*. Ed. Irving Howe (New York: McCall, 1970).

16. This charge made by Montiel Ortega may have been designed primarily to embarrass, COPEI Interior Minister, Rafael Andrés Monte de Oca, who was the 'Hererrista' candidate for the President. *2001*, December 29, 1981, p. 7.

17. "Marcos Gómez: Puerta Aponte Está Equivocado," *Doble 6* 3 (December 4–10, 1981): 29.

6

THE MANAGERIAL IMPULSE

THE PLANNING SUBSYSTEM

There are two basic factors underlying the introduction of planning as a key political subsystem in Venezuela and elsewhere. The first is the recognition that effective public policy in the twentieth century requires accurate factual knowledge of a given problem and the reasonable compliance of crucial nongovernmental groups in implementing public policy. The art of planning is the art of gaining this knowledge and support. Contemporary democracy must have a functional as well as representational component. The analogy presented by Robert Dahl that expertise, as well as popularity, must be given its due is appropriate.[1]

Pluralist societies and pluralist democracies are not mass societies. The aggregation of a winning electoral coalition is an obvious and central component of determining public policy. Yet, while the political impulse determines the direction in which a democracy moves, a planning dialogue that properly manages this movement is by no means antidemocratic. This chapter views the blending of the political and managerial impulses in Venezuela's pluralist politics.

In Venezuela, a most critical factor regarding planning has been that this subsystem has been the arena in which the business community's managerial impulse has confronted the political impulse of the party system. In a very real sense, the horizontal inclusion of business-sector opinion into public policy by means of a planning dialogue has been the price that the business elites may have extracted from the system for going along with populist democracy.

The basic goal of the economic elite prior to the nationalization of the oil industry in 1976 was to depoliticize public policy regarding key economic issues. The business community sought to isolate and insulate economic policy making from party ideology and partisan conflict. While "not denying the importance of political parties in the life of a democracy," Venezuela's economic elites argue that "administration could be separated from politics."[2] The acceptance of electoral democracy and the political parties was followed by the insistence that the political parties could not "attribute to themselves the exclusive right to decide our destiny."[3]

This quotation is from a 1964 public statement of a group of top Venezuelan businessmen who organized the *Asociación Venezolana de Independientes* (AVI, Venezuelan Association of Independents) in 1962 as the business community's effort, in a time of extreme political crisis, to demonstrate its support of the Betancourt government. The AVI has been called "a major landmark in creating a positive linkage between private enterprise and democracy."[4] The immediate reason for AVI's public statement was its 1964 decision to support the Democratic Action-Democratic Republican Union (AD-URD) and National Democratic Front (FND) ample-based government under President Leoni.

AD's Raúl Leoni and his rural supporters and Uslar Pietri's urban political supporters who formed the FND following the 1963 elections were brought together by the AVI and the honest brokerage of URD's Jóvito Villalba. The politicians in 1964 were in retreat, and the managers were on the attack with regard to the formulation of key public policy. Planning, as a technically based system of dialogue, was possibly to replace politics as the critical arena of public policy making.

The notion of the functional representation of corporate associations in Venezuela first appeared in the 1936 constitution. It called for the creation of a national economic council made up of representatives of business, labor, the professions, and the universities to aid the president in a consultative manner in the formulation of development policy. Although a decree of the development minister created such a national economic council in 1946, it has never operated as a channel of functional representation. The real model for Venezuela's planning system is the post-World War II French experience with indicative planning and the popularization of this system in Latin America by the work of the United Nations Economic Commission for Latin America.

In late 1958, the provisional government established the Central Office of Coordination and Planning (CORDIPLAN) in the office of the presidency. In addition to planning the government's own development efforts by preparing periodic four-year plans, the 1958 decree-law charged CORDIPLAN with the task of harmonizing the private and public sectors' development activities. The top personnel of CORDIPLAN along with a number of key business leaders have long been the principal advocates of public and private sector collaboration in economic-development policy making.

Although the advocates of indicative planning in Venezuela have attempted to install a system similar to the French "concerted economy," they have adopted the label "democratic planning" because of its greater political appeal. Opponents in the political parties and on the left have insisted that "oligarchic" or "elitist" planning would be a more accurate description of this style of public policymaking. The essence of the Venezuelan planning subsystem is that public policy should result from a technically based, institutionalized negotiation or dialogue between the national executive and the representatives of the various concrete economic-interest associations. This style of politics emphasizes the application of technical knowledge and the participation of the important concrete economic forces in the formulation of public policy to the detriment of partisan politics and abstract principles.

The key distinction between the post-World War II European indicative planning system or "concerted economy" and that of the mid-1960s Venezuelan proposals was that while Europe's planning system consciously included labor unions and consciously sought close linkage with the elected parliaments, the Venezuelan proposal consciously sought to exclude both. The labor unions were excluded from the proposed planning dialogues ostensibly for two reasons. One was that the labor unions would be represented, indirectly, by the political party leaders (AD, COPEI, and URD) who had been elected to the government or the congress. The second was that the partisan-based labor unions lacked the technical capabilities to participate in the planning dialogue.[5]

The real reason for proposing a planning dialogue between the elected government and the economic elites rather than a more ample triangular consultation between the government, the business community, and labor was the intended depoliticization of key decisions regarding the future of the oil industry and the nature of Venezuela's efforts to industrialize. Using the perceived weakness of the Leoní administration, on one hand, and the perceived weakness of the Venezuelan oil industry, on the other, the business elites' planning offensive focused on securing the continued operation of the foreign or transnational oil companies in Venezuela into the 1980s.

Uslar Pietri in 1966, after ending the ample-based government experiment with AD and URD, publicly announced that "he had joined the government [in 1964] under the mistaken assumption that the new Minister of Mines [Manuel Pérez Guerrero] was prepared to reorient petroleum policy."[6] The key businessman who was to draft the "democratic planning" proposal in the 1965–66 period, Pedro Tinoco, Jr., was also to draft the 1966 petroleum policy of the peak business association, the Federation of Chambers of Industry and Commerce (FEDECAMARAS).

The basic strategy of the business community's managerial offensive in the 1964–67 period was to dislodge the party system and congress from their key positions in economic policy making. The keystone of the proposal—to have a technically based, depoliticized planning dialogue that would become the key arena for public policy—was the suggested creation of a government-private sector economic council. This model, presented in Figure 6.1, was not to be formally institutionalized, but it has been the theoretical underpinning of much of the reality of Venezuela's public policy making.

With regard to the oil industry, the business community and the transnational oil companies then actively involved in Venezuela acted through the Federation of Petroleum Industry Association created around 1969, which became one of the constituent members of FEDECAMARAS. The transnationals sought to impose the creation of a National Energy Council (*Consejo Nacional de Energía*) as the principal policy-making unit for the industry. The National Energy Council had in fact been created in 1959, ostensibly to have the greatest possible consultation prior to formulating national policy for the oil industry. Membership included the Bankers Association, FEDECAMARAS, the Venezuelan Labor Confederation (CTV), and universities as well as several key government ministries. It was never allowed to function as a depoliticized policy-making body despite the fact that the government promised to reactivate it in 1966.

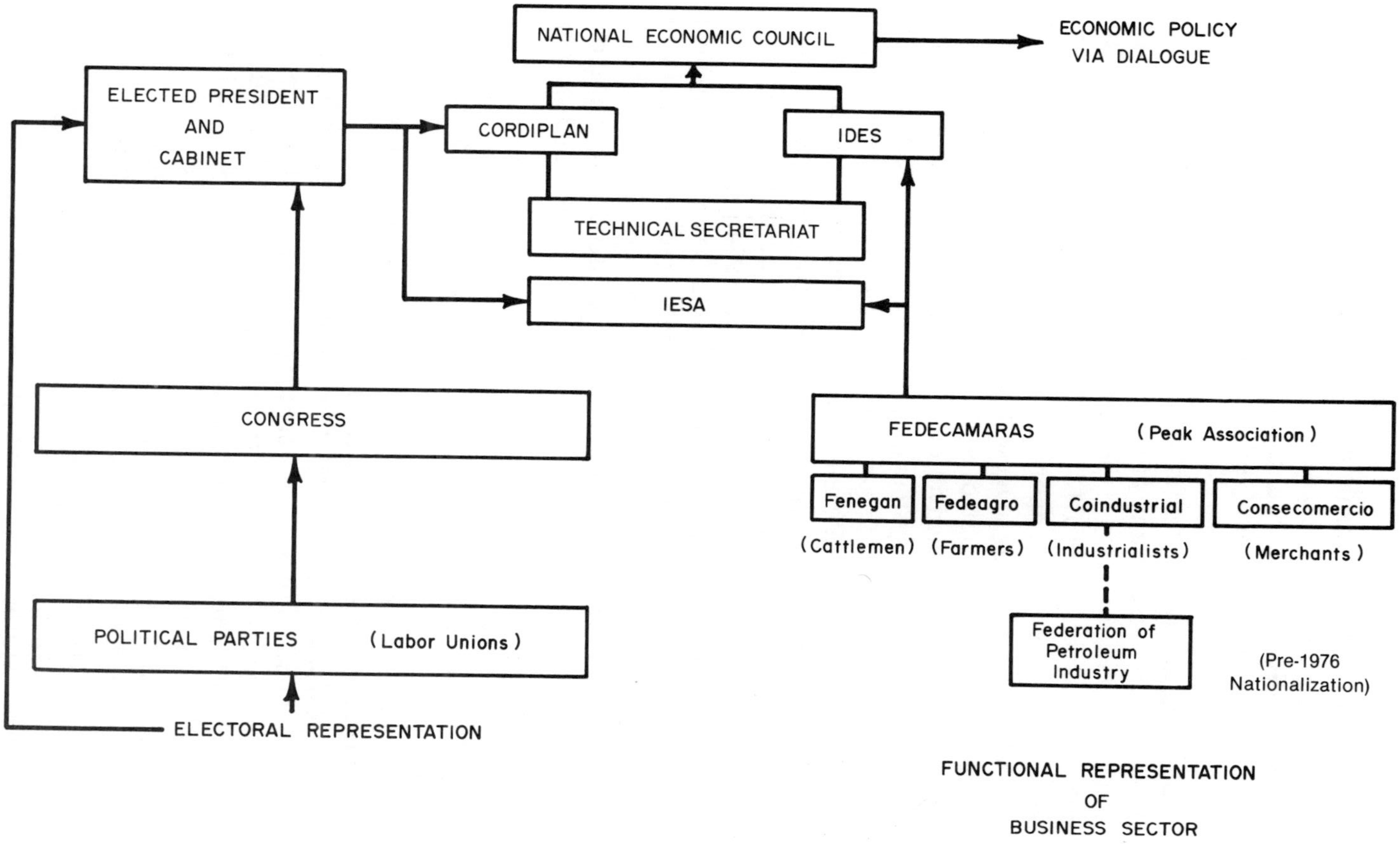

Figure 6.1. Schematic Representation of Depoliticized Economic Policy making by means of "Democratic" Planning

The interests of the transnational oil companies in activating the National Energy Council was based on the fact that they were a major factor in FEDECAMARAS and the Bankers Association. These Venezuelan surrogates would have been their representatives in the planning dialogue.

According to Tugwell, "gifts and dues from the petroleum industry far surpassed those of any other single contributor including the well-heeled banking industry;" the petroleum industry accounted for more than 10 percent of FEDECAMARAS's dues and "often made ad hoc contributions to support particular publicity campaigns."[7] There is strong evidence to suggest that this had been the case in the 1966 income-tax-reform controversy.

In pushing for the reactivation of the National Energy Council and of the comprehensive system of indicative or "democratic" planning, the transnational companies, led by Exxon, figured prominently among the 300 private companies that organized the Institute of Superior Administrative Studies (IESA) in order to provide a common graduate-level educational experience for future public- and private-sector managers. FEDECAMARAS in 1964 created its own planning unit, the Institute for Economic and Social Development (IDES). If the creation of IDES was to enable the private sector to fully participate in a broad policy determination with the government's CORDIPLAN, the creation of IESA intended to achieve a common managerial perspective among all future public- and private-sector managers.

The FEDECAMARAS and IDES analyses of the Venezuelan oil industry illustrate the business elites' goals. OPEC and Venezuela's role in the formation of that multinational commodity-exporting organization was hardly mentioned. The 1966 IDES report instead mentioned Venezuela's losing its relative importance in the international markets to Mideast competitors, citing studies that seemed to indicate that Persian Gulf oil producers were producing a barrel of oil at less than half the cost of a similar barrel produced in Venezuela. Calling the oil industry highly technical and vital to the nation because it had been the motor of its development, the IDES report recommended moderation in dealing with the transnational companies rather than confronting them with more tax claims and potential threats of nationalization.[8]

The business community, viewing the relative decline of Venezuela's position within the total world marketing of oil, the relative weakening of oil prices in the mid-1960s, and the progressive reduction of the nation's proven reserves, attempted to orient the government toward achieving two goals: reactivation of new foreign investment in the industry, and protection of Venezuela's share of the world petroleum market.[9] To best achieve these goals, a sense of partnership between the host nation, Venezuela, and its oil-producing guests, the transnationals, would be the preferred strategy. Confrontation with the companies was seen as being detrimental to the achievement of these goals. The business community suggested that a joint government-business effort to influence U.S. oil-importation programs seeking to include Venezuela in the U.S. system of "hemispheric preference," already granted to Canada and Mexico, should be the nation's primary concern regarding the industry in the late 1960s. The business community was also considering the enhanced "Venezuelanization" of the industry by means of creating several small privately owned Venezuelan oil companies, such as the Mito-Juan Oil Co. created in the 1960s.

The political fragmentation of the political-party system and increased uncertainty in the world's oil picture may have led to a considerable taming of government petroleum policy in the 1965–69 era, but it did not totally destroy it.[10] The achievement of self-determination and reasonably autonomous control over oil, its most vital natural resource, has been a constant goal of Venezuela's political impulse since the death of Gómez.[11] This value consensus and sense of historical hurt simply would not allow Venezuela's diverse and divided politicians to cogovern either with foreign oil companies or with the national business elite.

Despite heightened Venezuelan awareness of the vulnerability of its exports due to the prospect that a massive introduction of super tankers would lessen the transportation cost for Mideast oil and the fact that U.S. environmental controls were especially adversely affecting Venezuela's high-sulfur-content oil, the government proceeded to confront the transnational oil companies and the business community with a series of controversial reform proposals. These proposals concerned back taxes, production levels, controls, and a comprehensive income-tax reform.[12]

The Venezuelan domestic sector had one of the lowest tax responsibilities in the world, yet the early 1966 proposal by the Leoni government to moderately raise taxes and to moderately make the tax system more progressive shook the political system. The transnational oil companies and the domestic business community created a major political crisis and threatened a massive economic boycott of Venezuela. In some cases, the same businessmen who had just proposed a system of collaborative public- and private-sector planning dialogue that would do away with sterile ideological confrontations quickly mobilized massive resistance to the moderate 1966 income-tax-reform package.

The frontal assault of the managerial impulse in the 1964–67 period could not dislodge the political impulse from its central position in the political system, but the crisis it engendered in the late 1960s may have achieved a notable "back-door" victory. AD in 1967 fractured into two hostile parties, and in 1968, Rafael Caldera was elected president. Over the heated opposition of many in COPEI, President Caldera appointed Pedro Tinoco to be his minister of Finance. According to Tugwell, "Caldera's victory at first brought considerable satisfaction to the oil industry. Although the companies refrained from overt gestures in favor of any candidate, it was understood that they supported COPEI during the 1968 campaign."[13]

The Caldera administration (1969–74) was instead to become one identified with a period of confrontation and reform. During Caldera's administration, Venezuela cautiously reduced its dependency on both U.S. capital and U.S. markets by arranging government-to-government deals with a number of state-owned European oil companies.[14] The first step toward expanding Venezuela's oil reserves by beginning exploration of the Orinoco Heavy-Oil Belt (*La Faja*) started in the 1969–70 period.

Most importantly, a multiparty nationalist coalition led by Arturo Hernandez Grisanti of AD, Alvaro Silva Calderon (a disciple of Pérez Alfonzo) of the People's Electoral Movement (MEP), and Jesús Bernardoni of the Social Christian Party (COPEI) pushed through the Hydrocarbons Revision Law in 1971.

All this was a prelude to the 1973 oil shock caused by the boycott of the West by the Arab oil-exporting nations and to the quadrupling of petroleum prices in 1974. Venezuela, with this new petrobonanza, moved to nationalize the transnational-oil-company holdings in 1975. A broad political consensus had developed in Venezuela that the era of the foreign ownership of the oil industry was at an end.[15] The nationalization of the foreign-oil concessions did not end the managerial impulse, it merely changed its goals and style of operations.

In addition to advocating the separation of administration from politics by means of planning, the business community also accepted its social responsibility regarding the peaceful resolution of a number of the nation's social problems such as education, community development, and health.

In 1956, the year that signaled an end to the petroleum boom and the year in which the Pérez Jiménez government sold new concessions to the oil companies, Exxon's Creole Oil moved to institutionalize its positive concern with Venezuelan problems. The company created *Fundación Creole* (Creole Foundation) as its philanthropic arm. Fundación Creole, along with the *Fundación Sears Roebuck de Venezuela* (founded in 1955) and *Fundación Shell* (founded in 1959), has played a significant role in developing Venezuela's educational system by financing projects ranging from Andrés Bello Catholic University to a number of vocational and technical secondary schools.

Worried about the prospects of a potential communist takeover in Venezuela or a bloody civil war, the Creole Petroleum Co. in 1961 moved to demonstrate its support of and confidence in the Betancourt government. At a time when many national businessmen were exporting their own capital, Creole created an industrial investment subsidiary, the Creole Investment Corp. (CIC). In its first year of operation, CIC helped to finance 16 new industrial and agricultural ventures—all divorced from the petroleum industry. In 10 of these ventures, CIC held only a minority equity with Venezuelan partners.

Of greater importance at this time was the Fundación Creole effort to organize the private sector's massive involvement in the resolution of the nation's social problems. Fundación Creole first undertook a survey of the existing philanthropic activities of Venezuela's 50 largest companies. The study revealed that the seven foreign-owned petroleum companies contributed 90 percent of the $2,500,000 private-sector contribution to social welfare. The indigenous business community's concern for its social responsibility during a time of national crisis was clearly inadequate. In response to these findings, Fundación Creole jointly sponsored a conference on the businessman's responsibility toward the community with the Fundación Mendoza and the Venezuelan Association of Executives. The conference led to the creation of the *Dividendo Voluntario para la Comunidad* (DIVIDENDO, Voluntary Dividends for the Community) in early 1964.

The requirement for joining DIVIDENDO had been that the company must donate from 2 percent to 5 percent of its profits to finance educational and social welfare programs. Since its foundation, DIVIDENDO has been active in funding private-sector educational, community development, and welfare projects. Its 1965 budget was about $17 million. Creole Petroleum Co. had considered DIVIDENDO to be a very useful tool in teaching average Venezuelans about the benefits of free enterprise and convincing them that they will have a brighter future under this system than under any other. Fundación Creole had not been reserved about the role it played in creating DIVIDENDO or about stating its objectives for doing so.[16]

Implicit in stimulating the social responsibility of the business community was the stimulation of its political involvement. Many of the men who organized the February 1963 conference on business responsibility for social progress were also involved in creating the Association of Independent Venezuelans, designed to bring the business community's economic resources to bear on the 1963 elections. One latent function of the 1963 conference was the decision of the native business community to ask both Fundación Creole and Fundación Shell to take up the slack of philanthropic contributions caused by their massive political-campaign contributions.

Although DIVIDENDO was modeled after the U.S. United Way campaign, the fact that a number of its directors insisted upon using that agency to project their own personal political philosophies and ambitions led to its decline in the 1970s. Despite the atrophy of DIVIDENDO, Venezuela's business community has continued to actively implement its belief in entrepreneurial social responsibility. In 1980, a number of businessmen concerned with the inability of government to reach all who should be reached designed and implemented a private-sector, nonprofit, adult-literacy program.

One of the novel managerial aspects of this program is that the adult who chooses to join this program is expected to purchase a literacy kit as a sign of the individual's sincere commitment to acquire literacy. The businessmen who manage this program believe strongly that Venezuela's poor must be willing to shoulder their share of the burden in their own "liberation" and should not expect to receive passively "paternalistic" government largess. In addition to actively sharing in the financial costs of the program, there is an element of local-level self-management. The program can be considered to be a capitalist version of individualistic "conscious-raising" of Venezuela's poor.

A CONCERTED ECONOMY OR MARRIAGE OF COMPLICITY (1974–80)

The 1976 nationalization did not herald the "liquidation of the Venezuelan petroleum industry" nor the liquidation of the political clout of the Venezuelan business community, as some managers publicly feared in the 1966–73 period. The stunning electoral victory of Carlos Andrés Pérez in 1973 and the petroleum windfall of the 1974–80 period exemplified all the paradoxes and promises of AD and Venezuela's populist democracy.

The electoral victory of Pérez was just about total. The 48.6 percent of the valid presidential vote Pérez won in 1973 gave his "democracy with energy" a clear mandate. AD gained 28 of the 49 Senate seats and 102 of the 203 Chamber of Deputies seats. The congress, which fought Presidents Betancourt, Leoní, and Caldera to near distraction, could now become an instrument of an energetic implementation of the 1973 electoral mandate. In addition, the leadership of Venezuela's Confederation of Labor was, once again, in the hands of AD politicians.

AD in 1974 also contained an experienced team of top-level public managers or "politechnicians," a term coined by Hector Hurtado, a top AD economic planner, to suggest an individual whose career merges both political abilities and professional competence. Hurtado was named minister of finance in 1974. The man named to be the government's principal planner and director of CORDIPLAN, Gumersindo Rodríguez, was another respected politechnician. As a young AD leader, Gumersindo briefly left the party in 1960 to become a Revolutionary Left Movement (MIR) economic spokesman. He, however, quickly rejected MIR's drift toward violence and returned to AD after a number of years spent acquiring a British education. His rhetoric in outlining the government's economic plan in 1974 was pure European social democracy.

President Pérez had the option of attempting a model of democratic reform.[17] Instead, he opted for a "back-door association" of his government with a group of key entrepreneurs who have since been pejoratively labeled "the 12 Apostles."[18] The key to his opting for a model-state capitalism in association with these businessmen was his decision in April and May of 1974 to insist that Venezuela's elected congress grant him special "extraordinary economic and financial powers." This was done in May 1974, enabling Pérez to govern by decree rather than in conjunction with congress and its legislative and oversign functions. The core of President Pérez' extraordinary powers was the creation of three autonomous development funds whose management was not to be subjected to congressional oversight: the Venezuelan Investment Fund, the Industrial Investment Fund, and the Agricultural Investment Fund.

The primary purpose of creating these depoliticized or politically sanitized funds was to transfer the extraordinary government revenues resulting from the 1973 oil shock into "productive investment in the private sector."[19] The Venezuelan Investment Fund (FIV) had an additional purpose: to "sterilize a part of the additional oil revenue" by exporting capital out of the country in order to both diminish inflation and build up foreign-exchange reserves for future use.[20]

In addition to joint presidential-private sector management of these three funds, Pedro Tinoco, Jr. was named by President Pérez to head the Presidential Commission for the Comprehensive Reform of Public Administration. This commission focused on the question of providing efficient management for the myriad of state enterprises and joint state-private sector enterprises that have become an integral feature of Venezuela's contemporary state capitalism, or the "entrepreneurial state." The complexity of the nature of Venezuela's statist economy can be illustrated by the fact there are an estimated 226 public enterprises in Venezuela taking the form of state-owned companies, mixed enterprises (in which the state capital is less than 50 percent of the total), and autonomous institutions.[21] A list of the 15 key public enterprises comprises Table 6.1.[22]

Table 6.1. Fifteen Top Public Enterprises, 1979

Ranking Among Top 100 Enterprises[a]	*Type of Enterprise*[b,c]	*Name*	*Industry*
1	State enterprise	Petroleum of Venezuela (PDVSA)	Oil
2	Autonomous institute	National Institute of Sanitary Works (INOS)	Sewer services
3	State enterprise	Orinoco Steel Co. (SIDOR)	Steel

5	Autonomous institute	Venezuelan Social Security Institute (IVSS)	Health, hospitals
6	State enterprise	National Telephone Co. (CANTV)	Telephones
8	State enterprise	National Electric Corp. (CADAFE)	Electricity
12	State enterprise	Venezuelan Navigation	Merchant marines
14	State enterprise	Venezuelan International Airlines (VIASA)	Airline
15	Autonomous institute	Agricultural Marketing Corp. (CorpoMercadeo)	Distributor of agriculture
24	State enterprise	Venezuelan Aluminum Co. (Venalum)	Aluminum
25	State enterprise	Orinoco Mining	Iron mining
27	State enterprise	Aluminum Co. of Caroni (ALCASA)	Aluminum
28	Autonomous institute	National Housing Institute (INAVIL)	Public housing
29	Autonomous institute	National Institute of Ports	Port management
31	State enterprise	Venezuelan Dairy Industry	Dairy products

[a]1979, according to income, *not* profits or employment.

[b]A state enterprise has at least 50 percent of its capital owned by the government.

[c]Autonomous institutes generally provide basic public services such as water, telephone, and sanitation. There are user-fees charged for these public utilities.

Source: "Las Empresas Públicas," *Numero Revista de Economia y Negocios* 1 (August 1980), p. 17.

Prior to 1973, the politics of economic planning in Venezuela largely focused on capturing resources from the foreign-owned oil sector. Industrial planning in the nonoil sector had been largely limited to import substitution. Pre-1973 exceptions were the state-owned steel industry of the Orinoco (SIDOR), part of the Venezuelan Guayana Corp. (CVG), and the Petrochemical Institute. The nationalization of iron and petroleum and the petrobonanza of 1973 changed the principal role of the state from that of a "tax collector on primary-export products to direct production of primary and industrial goods."[23]

The V National Plan: 1976–80 drafted by the Pérez administration called for a total public- and private-sector investment of $53 billion, an amount more than three times that spent during the Caldera administration.[24] There was a fear in 1974 that the petrobonanza would be short-lived and, more critically, that Venezuela's oil reserves could be depleted by the late 1980s. The Pérez government argued that the 1975–80 period could be Venezuela's last chance to stimulate economic diversification and to consolidate electoral democracy. This pressing national need could have justified the installation of an institutionalized "concerted economy" in which the business community, with its managerial know-how, could have shared public-policy responsibility with the Pérez government by means of a planning dialogue.

This private business-government cogovernment, however, did not take place. According to scholars Coronil and Skurski, the Pérez style of government was a "marriage of complicity between Pérez's political clique and his economic clique, between 'bourgeois politicians' and 'political capitalists.' "[25] According to this thesis, Pérez represented a clique of politicians and businessmen who first secured control of AD following the MEP defection and secondly secured the election of Pérez. While this judgment of the events of 1974–79 may be too harsh, there is evidence to indicate that the Pérez administration represented a traditional caudillo-style takeover of the government. The 12 apostles associated with Pérez apparently acted in an individualistic and particularistic manner.[26]

A number of distinguished Venezuelan businessmen such as Eloy Anzola Montauban went on record opposing Pérez's insistence on extraordinary powers and governing by decree. According to Montauban, "Congress could not be ignored in a democracy."[27] A second critic in the business sector, Garcia Araujo, also suggested that the management of the new and growing public enterprises constituted a "new oligarchy."[28] Not only were they exempt from congressional scrutiny, according to this critic, but also their management of public funds was not subjected to the standard accounting practices required by the Commercial Code.

The growth of these massive public enterprises necessitated a new system of democratic oversight in order that the worst fear of Garcia Araujo, that their administration would constitute a "new oligarchy," not be realized. The populist political parties seemed to have failed to provide such an oversight, preferring rather to take part in the distribution of spoils and the allocation of featherbedding positions.

The Presidential Commission for the Comprehensive Reform of Public Administration made two key suggestions to restructure the oversight of the public enterprises.[29] One was to clearly distinguish between public enterprises, such as Venezuelan International Airlines, Orinoco Steel Co., and National Electric Corp. that were expected to make a profit and those, such as the Social Security Institute and Institute of Sanitary Works, that primarily were designed to provide a social service.

The second proposal, heatedly contested by the populist political parties, was to institute a system of "double inscription" of the state enterprises. In addition to being inscribed to a ministry, the Reform Commission proposed to inscribe related public enterprises to "superholding" sectoral corporations and to a new ministry for public enterprises. This superholding-company model already existed in the case of the Venezuelan Guayana Corporation and was the model for the creation of Petroleum of Venezuela. The fear of concentrating power in the hands of a supermanager or a council of sectoral corporations and the fear of a cogovernment with business managers led congress to table consideration of this basic reform. In 1982, it appeared that the Central Bank and Ministry of Finance had moved to become a sort of holding-company controller of these state enterprises.

THE MANAGERIAL IMPULSE IN CONTEMPORARY VENEZUELA

The managerial impulse, as distinct from the traditional and particularistic marriage of complicity, intervened to try to bring some order to the sprawling state intervention in economic production. One device was the creation of mixed enterprises in which state participation would be less than a majority share. The mixed enterprise would amalgamate the externalities of the public enterprise (concern for society, not just profits) with the private-enterprise goal of efficient management. For example, the state enterprise Orinoco Steel Co. (*SIDOR*) had the "best calibre management" in the public sector, yet it could not consistently make a profit. (In the mid-1970s, it started to show a slight profit prior to being disrupted by its rapid expansion under the Pérez administration.) *SIVENSA,* a smaller, privately owned steel firm, was making a profit despite the fact that the rules of the game had been stacked in favor of SIDOR.[30]

The public-enterprise managers of SIDOR were quoted in 1977 as suggesting that the imposition of redundant workers and the law making dismissal of workers quite difficult prevented efficient operation. SIDOR was the showcase steel plant in the frontier, planned city of Ciudad Guayana, located on the Orinoco River. Public policy and concern that the large unemployment existing in this frontier community could lead to massive rioting was one cause for SIDOR's policy of overemployment. SIVENSA was not placed under the same pressure to hire available warm bodies. In the late 1970s, SIDOR and SIVENSA merged part of their operations to create a new mixed enterprise, FIOR of Venezuela. SIDOR also formed a mixed enterprise with the French firm *Nobel Bozel.*

Successful mixed enterprises were created in the critical aluminum industry and in the paper and pulp industry. According to Gene Bigler, of the 293 public entities (a concept somewhat different from that of a public enterprise), 123 had some private participation in their capital in the late 1970s.[31] The two sectors with no private capital were petroleum and public utilities (gas, water, electricity).

The second managerial device to control sprawling public enterprises, the fear of a new oligarchy, and to prevent abuses by the traditionalist marriage of complicity was the much-maligned foreign investment. European scholar Fred Jongkind "pointed to the danger that the government's aspiration to reduce foreign dependence could result in another form of dependence, namely domination of the market by [domestic] monopolies or oligopolies."[32] In a sense, the "intense intrabourgeois conflict" Coronil and Skurski mentioned in their automotive-industry study, allegedly between "ascendant" and "dominant" family-based economic groups or between foreign and national businesses, is one device to preserve not only an efficient market but also political pluralism.[33] The concentration of economic power has always been a problem for democracy. Having diversified and competitive forms of economic organization prevents this overconcentration.

Foreign investment properly guided by a democratic state, as it appears to be in Venezuela, could make a positive contribution to the nation's autonomous development. In addition to the Andean Pact

Decision 24, which establishes a multinational Code of Conduct, Venezuela has created its own Superintendency of Foreign Investment to encourage foreign investment under conditions favorable to Venezuela. The role of foreign investment as a device to prevent monopolistic or oligopolistic control of industry was one proposal of the second Merhav report.

The Hererra administration in 1979, as part of its initial effort to gain control of the sprawling public sector, hired Israeli economist Meir Merhav once again to study the Venezuelan economy. The second Merhav report found that the economic organization of Venezuela combined the negative element of central government planning and economic intervention "with the negative element of laissez-faire,"[34] in other words, the worst elements of both socialism and capitalism without the redeeming features of either system. The report's own version of the marriage of complicity between certain politicians and certain businessmen follows.

The populist politicians enacted rigid tariff protection, price controls, and public-credit criterion. The businessmen associated with these politicians were granted protection from foreign competition, privileged access to government credit, and immunity from certain regulations. The private business sector ended up more dominated by inefficient monopolies and/or oligopolies, while the sprawling public enterprises were free from congressional oversight and the discipline of the market. Neither the public enterprises nor the protected private enterprises were required to produce quality products at competitive prices.

According to the 1982 report of the comptroller-general of Venezuela, the constitutionally authorized watchdog of public spending, the *insouciance,* or nonchalance, with which the 226 public enterprises were managed caused an apparent loss of approximately $8 billion in a two- or three-year period.[35] Productivity in industry fell 5 percent during the 1974–77 period, when growth should have been expected, and the public debt reached an astronomical amount of $30 billion in 1983. Inflation also reached 21 percent in 1980 as the economy ground to a halt.

The Merhav report's biggest criticism concerned the state's provision of human services. Despite more than 20 years of populist democracy, a study of 112,000 Venezuelan children revealed a nation that was largely "undernourished, undereducated, and often physically weak."[36] It appears that for nearly 20 years, the populist politicians attacked the nation's social problems by spending money on rapidly built, tangible, capital projects rather than on developing sound programs to improve the nation's health, education, and welfare.

The noticeable deterioration of basic public services, the massive mismanagement of public funds, and the continuing growth of inequality was the end result of the petrobonanza. After spending $50 billion, Venezuela showed a muddled economy and an inefficient and inequitable society rather than a mixed economy and social reform. The Venezuelan public's loss of faith in the ongoing democratic experiment is due in large measure to the calamity of the "enterpreneurial state."

Mention should be made of the continuing, institutionalized, economic-oversight function provided by peak business associations such as FEDECAMARAS and the more nationalist Pro-Venezuela Association. These business associations have held forums on key development issues, pointing out implicitly that public funds, perhaps controlled by the individualistic marriage of complicity previously mentioned, may not have gone to the most deserving businessmen or companies. A second theme that is a constant irritant to the businessman is the constant expansion of petty government regulations that do not really regulate but become excuses for additional bribes. Institutionalized business associations have become key champions of those in the public sector who have sought to clean up the public administration.

The President of the Central Bank of Venezuela, Leopoldo Díaz Bruzual, has emerged as an activist proponent of the managerial impulse. This Swiss-educated Christian Democrat served President Caldera as president of the Foreign Trade Institute and President Hererra as president of the Venezuelan Investment Fund prior to his being "retired" to the Central Bank. According to Díaz Bruzual, the growing state intervention in Venezuela was not the result of any ideology but more correctly "the result of an erratic and improvised process."[37]

According to Díaz Bruzual, the oil boom of the mid-1970s accentuated the power of the already established, and allegedly inefficient, private economic groups, which refused to take risks or to be enthusiastic about taking on new ventures. The state, because of this failing on the part of the dominant economic groups, was obliged to do more than it could competently handle. This, according to Díaz Bruzual, is the origin of the present collapse of basic public services.

The criticisms of Díaz Bruzual have been criticisms of Venezuela's ongoing populism: that the state simply does not know how to authentically redistribute national income. The massive expansion of public employment as the populist alternative to democratic reform has been a tragedy. This army of poorly educated and poorly paid public employees has responded to its own plight by low productivity and creative corruption.

The managerial impulse is not necessarily antielectoral democracy; it is antipopulist. It has energetically criticized the reality, under populism, of the interface of government regulations and private-business behavior. Rather than truly achieving a public purpose, there is evidence to suggest that government officials and bureaucrats charged with overseeing the implementation of these regulations do so with private gain in mind. There is a pejorative term used in Venezuela to describe regulations that do not regulate, but only encourage bribery; *permisologia*. The art of permisologia makes official permits and approval vital parts of business.[38] The powerful business groups with significant political connections are in fact free of this harassment. The second Merhav report suggested that the operations of permisologia enabled these dominant groups to maintain their oligopolistic control over Venezuela's economy.

NOTES

1. Robert Dahl, *After the Revolution: Authority In A Good Society* (New Haven, Conn.: Yale University Press, 1970).

2. "AVI Pronuncia a Favor del Jobierno de ample Base," *El Nacional*, July 14, 1964. For a detailed discussion of the development of the Venezuelan business community's planning ideology, see David E. Blank, "Policy-Making Style and Political Development: The Introduction of a System of Democratic Planning in Venezuela 1958–1968," (Ph.D. dissertation, Columbia University, 1969), and David E. Blank, "Political Conflict and Industrial Planning in Venezuela," in *Venezuela 1969: Analysis of Progress*, ed. Phillip B. Taylor, Jr. (Houston: Office of International Affairs, University of Houston, 1971).

3. Ibid.

4. José Antonio Gil Yepes, "Entrepreneurs and Regime Consolidation," in *Venezuela: The Democratic Experience*, ed. Juan Martz and David Myers (New York: Praeger, 1977), p. 146.

5. The democratic planning concept in the French context had an entirely different meaning than that intended by its Venezuelan proponents. For an understanding of the French system, see Pierre Mendes-France, *A Modern French Republic* (New York: Hill and Wang, 1963), and John Hackett and Anne-Marie Hackett, *Economic Planning in France* (Cambridge, Mass.: Harvard University Press, 1963).

6. Franklin Tugwell, *The Politics of Oil in Venezuela* (Stanford, Calif.: Stanford University Press, 1975), p. 86.

7. Ibid., p. 99.

8. Pedro Tinoco, R., ed., *Analisis de la Política Petrolera Venezolana* (Caracas: FEDECAMARAS, 1966), and Pedro Tinoco, R., ed. *FEDECAMARAS Planci-ficación Democratica* (Caracas: FEDECAMARAS, 1965). Tinoco in the 1960s was a director of the Chase Manhattan Bank's Venezuela subsidiary and an advisor to the foreign oil companies and to IDES.

9. Humberto Peñaloza, "The Political Framework of Venezuelan Oil: Changes and Opportunities," in *Venezuela 1969: Analysis of Progress*, ed. Phillip B. Taylor, Jr. (Houston: Office of International Affairs, University of Houston, 1971), pp. 197–215.

10. Tugwell, *The Politics of Oil*, p. 99.

11. John D. Martz, "Policy-Making and the Quest for Consensus: Nationalizing Venezuelan Petroleum," *Journal of Inter-American Studies and World Affairs* 19 (November 1977); 485.

12. Tugwell, *The Politics of Oil*, pp. 86–88.

13. Ibid., p. 100.

14. Ibid., pp. 118–19.

15. Martz, "Policy-Making and the Quest," p. 485.

16. Leo E. Lowry, president of Creole Petroleum Co., "Algunos Aspectos de la Imagen de la Empresa Norteamericano en Venezuela," in *Venezuela, Panorama 1969: Una Mirada Al Futuro* (Caracas: Creole Petroleum Co., 1970), pp. 28–29. For a number of reasons, not all of which are clear to this author, the above Spanish-language version is not identical to Phillip B. Taylor's English version, *Venezuela 1969: Analysis of Progress*, Lowry's 1969 presentation made in Washington D.C. is not included in the English-language version.

17. An outline of the potential for substantial democratic reform leading to greater equity in the distribution of national wealth and full employment, as well as the feared big business-big government cogovernment, appear in David E. Blank, "The Politics of Industrial Planning in Venezuela 1958–1974: Variations on the theme of 'Democratic Planning'," (Paper presented at the fifth national meeting of the Latin American Studies Association, San Francisco, November 14–16, 1974). See also *Semana Confidencial*, May 9–15, 1974, and "¿Socialismo O Capitalismo?," *Elite* (June 1974).

18. The origin of the expression "the 12 Apostles" to describe the businessmen collaborators of President Pérez was apparently Pedro Duno, *Los Doce Apostoles* (Caracas: Editoral Badell Hermanos, 1976).

19. Pedro-Pablo Kuczynski, "The Economic Development of Venezuela: A Summary View as of 1975–1976," in *Contemporary Venezuela and Its Role in International Affairs*, ed. Robert D. Bond (New York: New York University Press, 1977), pp. 56–57.

20. Ibid.

21. Allan R. Brewer-Cariás, *Régimen Jurídica de las Empresas Públicas en Venezuela* (Caracas: Editorial Arte, 1980).

22. "Las Empresas Públicas," *Numero: Revista de Economia y Negocios* 1 (August 1980), pp. 14–21.

23. Fernando Coronil and Julie Skurski, "Reproducing Dependency: Auto Industry Policy and Petrodollar Circulation in Venezuela," *International Organization* 36 (Winter 1982), p. 60.

24. Mauricio Garcia Araujo, "The Impact of Petrodollars on the Economy and the Public Sector of Venezuela," (Paper presented at the tenth meeting of the Latin American Studies Association, Washington, D.C., March 4, 1982), p. 5.

25. Coronil and Skurski, "Reproducing Dependency," p. 70.

26. Luis Lander, "Prologo: Proceso Político," in *CAP 5 Años: Un Juicio Critico* (Caracas: Ateneo de Caracas, 1978).

27. Eloy Anzola Montauban, "A Voice of Opposition: The Executive has Ignored Congress," in Caracas *Daily Journal*, May 24, 1974.

28. This quote of Garcia Araujo appears in José Antonio Gil Yepes, *El Reto de las Elites* (Madrid: Tecnos, 1978), p. 84.

29. Ibid., pp. 88–90.

30. Juan Pablo Pérez Sainz and Paul Zarembka, "Accumulation and the State in Venezuelan Industrialization," *Latin American Perspectives* 6 (Summer 1979): 22.

31. Gene Bigler, "The Decentralized Public Administration and Venezuelan Political Economic Development," (Ph.D. dissertation, Johns Hopkins University, 1978), quoted in Gil Yepes, *El Rato de las Elites*, p. 91.

32. Fred Jongkind, "Industrialists and the State: Private Views on the Public Role in the Industrialization of Venezuela," (Paper read at the national meeting of Latin American Studies Association, Pittsburgh, April 5–8, 1979), p. 12.

33. Coronil and Skurski, "Reproducing Dependency," pp. 70–71.

34. Joseph Mann, Jr., "Merhav's Venezuela: Poor Little Rich Country," *Business Venezuela* (July/August 1980): 36.

35. *Latin American Weekly Report*, May 28, 1982, p. 3.

36. Mann, "Merhav's Venezuela," p. 24.

37. Díaz Bruzual, "El Estado no Supo Redistribuir," *Numero: Revista de Economia y Negocios* (August 1980): 118–19.

38. Joseph A. Mann, Jr., "Halfway to Nowhere," *Business Venezuela* (May/June 1981): 16, and "Venezuela's Voracious Bureaucracy," *Business Venezuela* (July/August, 1982), pp. 14–20.

7

MANAGEMENT AND POLITICS: THE CASE OF OIL (1904–76)

THE AGE OF THE TRANSNATIONALS (1904-36)

The traditional *caudillo* (warlord), Cipriano Castro, sought foreign oil investment as a means to gain sufficient funds to defeat domestic rivals and to prevent the threat of European military intervention like that which occurred in 1903. U.S. President Theodore Roosevelt may have considered U.S. involvement as news of potential oil in Venezuela spread.[1] The start of the Mexican Revolution in 1910 and the replacement of Castro by Juan Vicente Gómez in 1908 opened Venezuela to foreign-oil-company investment. The strategic needs of World War I encouraged the British- and Dutch-owned Shell Co. to produce the first commercially successful well in 1917. Massive U.S. oil penetration of Venezuela began in the 1920s, primarily through the efforts of Exxon (previously called ESSO, or Eastern Standard Oil of New Jersey) and Gulf Oil. By 1928, oil had become Venezuela's principal export and was on its way to transforming this once-impoverished nation.

Gómez, always the clever caudillo, tried to limit the impact of oil on his nation for fear of secession, as in Panama, developing in the oil rich state of Zulia, where Lake Maracaibo and its early petroleum discoveries were located. Gómez never allowed foreign companies to dredge the sandbar that prevented ocean-going ships from entering the lake. He reasoned that if gunboats could enter the lake, foreign intrigues to have Zulia secede from Venezuela would quickly follow.

For similar reasons, Gómez did not press Shell or Exxon to refine Venezuelan oil in Venezuela. The Dutch Islands of Aruba, Curacao, and Bonaire were conveniently close to the western coast of Venezuela, so Shell and Exxon built their refineries 35 miles off shore.[2] Gómez's discouragement of domestic refining for reasons of security denied Venezuela access to the acquisition of skills, employment, and revenue.

The oil camps developed by the foreign companies on their concessions were also allowed to be isolated, fenced-in foreign enclaves, or states within a state. Since much of the early oil was discovered in the malaria-infested swamp of Lake Maracaibo, native Venezuelan workers initially were not excited about seeking jobs there. Perhaps reflecting both government and foreign-company decisions to maintain a divided and difficult-to-organize labor force in these oil camps, West Indian, English-speaking, black laborers were brought in.

As the companies moved to drain the swamps to control malaria and to provide basic public services, the influx of native Venezuelan workers increased. The tension between foreign and domestic laborers led to many clashes and apparently prevented any unified labor movement from emerging in the 1920s. In 1929, after the beginning of the world depression, Gómez moved, first to stir up racial resentment and second to bar the entrance of additional black workers.[3] About 23 percent of the labor force active in the oil industry at that time was foreign.

Descriptions of the oil-boom towns that sprung up along Lake Maracaibo were all quite romantic. One contemporary U.S. account of Lagunillas, one of the boom towns of the 1920s, speculated on the confusion a future archeologist would have in describing the "asphalt culture" of the 1920s:

> He'll be puzzled by wandering parts of oil-field machinery. He'll find a phonograph next to a dug-out canoe. He may find a Chinese opium pipe and a Syrian narghile. . . . He will undoubtedly find many contraceptive devices, and perhaps will argue that it was deliberate race suicide which extinguished the "asphalt culture." But a hated rival is sure to turn up a few preserved babies and prove it was infanticide which did the trick. . . . He'll find a pair of Sears Roebuck cowhide boots if he looks long enough.[4]

These disorganized and spontaneously constructed oil towns were quickly to become the native quarters, often devoid of basic urban amenities. Close by were the fenced-in and patrolled oil-company camps, little self-contained versions of a Miami suburb or a British garden city. Good hospitals, good schools, and good living were to be found in these camps. The stark division of Venezuela into two separate worlds, of being and conceiving, had begun. Some Venezuelans, due to good fortune, connections, and skills, were admitted to the privileged world of the petroleum sector, but for the vast majority, there were the improvised shack towns and isolated peasant huts.

This stark social division was both cause and effect of the growing economic distortion of the economy. Simply put, if an unskilled worker in the oil camps earned one Venezuelan bolivar to do light work, why should the farm workers earn only a medio (a quarter of a bolivar) for long hours of picking coffee? Since the coffee economy could not afford to pay petroleum's wages, coffee did not get picked.

In the midst of a sea of corruption and caudillo indifference, the career of Gumersindo Torres, twice Gómez's minister of development, stood out as a beacon of Venezuela's nationalist response to the intrusion of oil. For the first time, official Venezuelan concern focused on getting a just share of the industry's profits and on fostering Venezuelan production of industrial inputs into the industry.[5] Gumersindo Torres established the Office of Hydrocarbons in the Ministry of Development in 1930, Venezuela's first official and institutional attempt to control oil. During the Great Depression, Torres began to charge foreign companies with flagrant abuse of Venezuelan law and demanded back taxes and reparations. Gómez sided with the foreign companies and fired Torres in 1932. According to Rómulo Betancourt, pressure from both the British and U.S. embassies had a hand in the firing of Torres.[6] However, the Torres ministry, the beginning of labor unrest in the oil fields, and fiscal devastation caused by the nation's almost complete dependence on oil income all indicated the passing of the era of caudillo-foreign company domination of Venezuela.

The death of Gómez in December 1935 signaled the violent birth of a new era. The scenes in Venezuela during the weeks following his death resembled the 1979 collapse of the Iranian monarchy. Rioters burned the Foreigners Club in Maracaibo, and the city center was sacked. A number of the fenced-in oil camps were surrounded by angry *mestizo* Venezuelan mobs, who took their frustrations out on some of the hapless black, West Indian workers they found when they failed to smash into the oil camps.[7] The Shell Co. evacuated its Lagunillas camp, sending foreign women and children to the relative safety of a tanker in the middle of Lake Maracaibo. The Gómez clan fled Venezuela as an estimated 25,000 marched on the Presidential Palace in Caracas.[8]

THE ROAD TO NATIONALIZATION (1936–76)

Starting with men such as Alberto Adriani and Uslar Pietri, advisers to Presidents López Contreras and Medina Angarita (1936–45), and the contributions made in the era of Juan Pablo Pérez Alfonzo (1945–48 and 1948–70), Venezuela slowly developed its goals and strategies for an independent response to oil. The intense, internal, political competition on oil matters notwithstanding, a broad-based national consensus was created at different times regarding the revolutionary achievement of the goals. Briefly, Venezuela was to focus on building domestic and international political support and institutional capabilities to achieve four goals:

> The achievement of a just share of the profits from the oil industry, which necessitated gaining increased control over management of the industry.
>
> The sowing of oil profits to sponsor an integrated national development and to overcome the divisions and distortions created by the intrusion of oil.
>
> The conservation of the nation's proven reserves to last until the end of the petroleum era.
>
> The creation of global conditions to secure stable markets for Venezuelan oil at a fair price.

Official Venezuelan government statistics, as indicated in Chapter 2, revealed that between 1919 and 1936, Venezuela received a mere 7 percent of the industry's profit from foreign concessionaires.[9] In 1936, few in Venezuela were going to tolerate the continuation of this situation. A combination of a bitter 43-day oil-workers' strike in late 1936 and the passage of the 1936 Labor Law started to improve the situation of the workers. The first fundamental reform of the nation's fiscal control over foreign companies, however, had to wait until the World War II period. Between 1936–40, the transnational companies were pushed, however reluctantly, to recognize that conditions had changed. Not even the Mexican oil nationalization of 1938 fully convinced the companies that the rules of the game had changed.

With the outbreak of World War II, the U.S. government intervened in the Venezuelan transnational-company negotiations, not only to implement its promise of being a good neighbor but also to secure the oil necessary to defeat fascism.[10] It was Venezuelan oil that enabled Great Britain to fight on in 1940. It required the direct intervention of Secretary of State Hull, the constant concern of the U.S. Embassy in Caracas, and a team of U.S. oil experts headed by Arthur Cartice and Herbert Hoover, Jr. (the son of the former president), hired by the Venezuelan government, to get the oil companies to agree to the 1943 hydrocarbon reform.[11]

The intense, democratic, political competition between the Democratic Action (AD) opposition and the Medina Angarita government "probably helped Medina strengthen his bargaining position with the companies."[12] The beginnings of democracy at home in the 1936–43 period and of the intimate collaboration between the governments of the United States and Venezuela during World War II created new rules for the politics of the oil game. Both Medina and AD were in agreement that their goal was "to obtain at least half the companies' future profits for Venezuela."[13] This goal was achieved by the AD-led revolutionary junta in 1946, and the institutionalization of the 50–50 principle spread throughout the world in the 1950s.

In time of war and postwar reconstruction, the transnational oil companies responded to Venezuelan offers that they and the nation proceed in the spirit of "partnership in a common enterprise."[14] Gulf Oil's Venezuelan subsidiary, *Mene Grande*, sent an agricultural mission to encourage the production of suitable food crops for its workers in the eastern Venezuelan oil fields. Both Gulf and Exxon agreed to pay back-tax claims, without much conflict, in the 1940s.[15]

Even with the imposition of the Pérez Jiménez dictatorship (1948–58), Venezuela refused to ease the 50–50 formula in its taxation of the industry, although there was a "gradual erosion" of the share of the industry's profits that went to Venezuela.[16] However, the Pérez Jiménez dictatorship continued the sowing-of-oil policy by massive investments in the nation's physical infrastructure, including highways, dams, and electrical plants. It even began construction of the promising Guayana region industrial development.

The 1957 decline in world oil prices, caused in part by the reopening of the Suez Canal and the expansion of world oil production, mixed with the free-spending style of the dictatorship's public finances to contribute to the collapse of Pérez Jiménez's rule in January 1958. The intense political atmosphere of 1958 caused the outgoing provisional government to raise the nation's share of the industry's profits to 65 percent. This sudden revision of the tax laws shocked the oil companies. In 1959, Juan Pablo Pérez Alfonzo again became the minister of mines.

The immediate problem facing Pérez Alfonzo in 1959 was the continuing decline in oil prices caused by the glut in production. This looming disaster was exaggerated by the March 10, 1958, decision of the Eisenhower administration to impose mandatory quotas on imported oil.[17] Despite the Eisenhower administration's desire to see the Betancourt government succeed, it refused to include Venezuela in the system of hemispheric preference granted to Canada and Mexico. The impact on Venezuelan oil exports to the United States by this quota was devastating. Its share of U.S. imports fell from 67 percent in 1957 to 42 percent in 1969.[18] Despite the fact that Venezuela was the keystone in the Kennedy administration's "Alliance for Progress" in the early 1960s, apparent transnational-oil-company pressure on the U.S. government and congress prevented a positive response to Venezuela's continual appeal to be included in the system of hemispheric preferences.

The ill-advised 1959 decision of the Eisenhower administration was the definitive milestone in the road to the Organization of Petroleum Exporting Countries (OPEC). In 1959, Pérez Alfonzo began his courtship of the Mideast oil producers by attending the first Arab Petroleum Congress in Cairo. Saudi Arabia's Abdullah Tariki joined in his effort. An August 1969 price cutback ordered by the transnational oil companies was the last straw. In September 1960, OPEC was established in Baghdad as four nations—Venezuela, Saudi Arabia, Iraq, and Iran, which together accounted for 80 percent of oil exports—moved to restore the price of their key natural resource.[19] The keystone of OPEC has been the producer nations' organized regulation of production levels as a means of controlling prices.

Pérez Alfonzo's long-term effort to instill a sense of common need and purpose among the diverse nations of OPEC started to pay off in 1969. The newly installed nationalist dictatorship in Libya requested and received a Venezuelan mission to assist it in regulating its oil industry. This Venezuelan mission, headed by Arévalo Guzmán Reyes, found irregularities in the transnational treatment of Libya.[20] Libya moved to raise its taxes and to cut production. This, in turn, helped push up world oil prices. With this global support, Venezuela broke out of the tamed attitude it had assumed toward transnationals at home in the mid-1960s. In 1970, the nationalist coalition in the Venezuelan Congress pushed through a price increase that reversed "a decade of decline."[21] Venezuela's 1970 action galvanized the Mideast members of OPEC. In 1973, the Arab nations (not OPEC) initiated their politically motivated oil boycott. The age of transnational ownership of oil was coming to an end.

It is worthwhile to review the second pillar of Pérez Alfonzo's oil policy. In 1960, two innovative domestic reforms of the oil industry were enacted into law: the policy of "no more concessions" and a state-owned oil company, *the Corporación Venezolana de Petróleo* (CVP, or Venezuelan Petroleum Corp.). The principal driving force in the management of the CVP between 1961 and 1967 was an experienced AD leader, Rubén Sader Pérez.

According to Sader Pérez, the twin objectives of the CVP were to enable the Venezuelan state to "directly intervene in all phases of the industry" (exploratory, production, marketing) and to assure to Venezuela an increased share of the industry's profits.[22] The CVP was not only organized to be a fully integrated oil company with profitable operations in mind but also to train Venezuelans to handle all phases of the industry. In 1966, the CVP was utilized to barter the oil it received from the transnationals as part of their royalty payments for Brazilian buses. This was designed in part to break the transnationals' monopoly over the international marketing of oil. The CVP was also to be given a monopoly of all marketing of gasoline within Venezuela. Finally, the CVP, by initiating direct government-to-government exchanges of oil for technology, started the subtle and gradual erosion of transnational control over the global distribution of technology.

According to Sader Pérez, this Venezuelan "entrance to the exterior"—the international market—by means of "barter" preserved the principle of defending the price of oil.[23] The most important global breakthrough of the CVP was the initiation in 1966 of the policy relating to the technologically advanced transnational oil companies by means of service contracts. This linkage of the state-owned

CVP with the giant transnationals—Exxon, Shell, Gulf, and Texaco—was begun in the "climate of mutual understanding" that existed in the late 1960s after a rather intense confrontation between Venezuela and the companies over the 1966 income-tax-reform proposals.[24] These service contracts, formally initiated in 1967, offered the possibilities to the CVP and the different transnationals to form mixed companies (of a public-private or joint ownership nature) and to develop innovative technology jointly.

The establishment of a service-contract relationship between the CVP and the transnationals in 1967 and the income-tax breakthrough achieved by Venezuela in 1970 were developments of the utmost importance for two reasons. The implications of the cumulative tax increases on the transnational companies was that between 1968 and 1973, the effective share of profits that would go to Venezuela would rise from about 72 percent to 75 percent.[25] At this projected division, the transnationals would have little incentive to maintain any sort of sophisticated investment or long-term commitment to Venezuela. In fact, there is little doubt that they consciously disinvested in Venezuela from 1959 onward. While professing to be good partners of Venezuela, there is reason to believe that, given the climate of confrontation that existed in the 1959–66 period, the transnationals "never forgave Venezuela for the tax increase at the end of 1958."[26]

This charge against the transnationals was not made only by liberal-minded U.S. academics. In 1977, the late President Betancourt wrote, "In 1959, the companies began a well coordinated plan of disinvestment in Venezuela, and at the same time, they intensified their explorations in Canada and in the Middle East."[27] Betancourt proceeded to quote Tugwell's "revealing" 1975 book, suggesting that the 1959 creation of the Petroleum Chamber of Industry was part of the transnationals' strategy, publicly proclaimed by Creole's former director, Harold Haight, "to force the Venezuelan Government to lower the taxes and open tenders for new concessions." Betancourt concluded his favorable comments on Tugwell's book by quoting the financial contributions the transnationals had made to FEDECAMARAS.[28] One impact of this disinvestment was that employment in Venezuela's oil industry fell 28 percent between 1960 and 1966.

The second reason for working out the service-contract system was that the politics of confrontation (1959–66) and the tension they generated had exhausted their possibilities. Venezuela had gone as far as it could in increasing its participation in both the industry's decision-making process and its profits. The time was propitious for a change to the managerial approach with its focus on relaxed, mutual achievement of efficient profits for the nation and the transnationals: the art of having all winners and no losers. The service contracts were the managerial means of achieving this new level of Venezuelan-transnational understanding and collaboration.

The time to consider the ultimate question of the ownership of the industry was at hand. There was too much uncertainty in both the domestic and global politics of oil to assure that the establishment of a service-contract relationship between the CVP and the transnationals would secure the future. Among other reasons, the CVP in 1970 was still a rather minor operation, accounting for only 2 percent of the nation's oil production. The fact that 14 of the transnationals began negotiating with it in 1967 to begin exploration of two new areas—the Gulf of Venezuela (an off-shore development in waters still disputed with Colombia) and the zone of South Lake Maracaibo—indicated that Venezuela was focusing its priority on finding new oil.

Suspicion on both sides would have probably prevented any long-term service-contract relationship. Indicative of these suspicions was the Venezuelan charge that behind the facade of being Venezuela's partner, Exxon had in fact firmly opposed any preferential treatment for its host's oil with regard to Venezuela's long-standing concern of being granted hemispheric preference in the U.S. market.[29]

The political impulse's response to this reality was the congressional imposition of the 1971 Hydrocarbons Reversion Law. This law called for the incremental reversion to the nation of all existing concessions upon their expiration, starting in 1983. The law further stipulated, in response to the industry's disinvestment, that unexploited territorial concessions would be ceded to the CVP in 1974. In addition, the transnationals would be required to post bonds in order to guarantee that their capital investment in the form of plant and machinery would be turned over to Venezuela in good working condition.[30] It should be remembered at this juncture that Venezuelan law was based on the colonial royal-patrimony approach to subsoil wealth, and the government traditionally insisted that the oil still in the ground always belonged to the nation. The royal patrimony had become the national patrimony.

The passage of the 1971 Reversion Law was denounced as being confiscatory and a de facto nationalization. The effort by both the nation and the transnationals to proceed in a "climate of mutual understanding" was threatened. Three conservative U.S. senators with close ties to the oil industry tried to get the Nixon administration to prohibit imports from Venezuela in an effort to block the passage of the bill.[31] This absurdity was matched by the seizure of a U.S. engineer by Venezuelan border guards prior to crossing into Colombia with laboratory equipment and papers the Venezuelan guards insisted were national patrimony under the 1971 Reversion Law.

The global aspects of the politics of oil in the 1969–71 period, immediately prior to the October 1973 Mideast and Arab oil embargo, contained two critical components for Venezuela. The decade-long Venezuelan effort to receive hemispheric preference was reaching a climax at a time when Richard Nixon assumed the U.S. presidency. Venezuela was reluctant to accept Nixon's first choice for U.S. ambassador to Caracas, John G. Hurd, a member of the Independent Petroleum Producers Association, which was the principal lobby for excluding Venezuelan oil.[32] His nomination was like offering to send a Zionist to Cairo. It is noteworthy that in 1958, Vice-president Nixon felt that his life was threatened in Caracas, and in 1969, Vice-president Rockefeller cancelled his visit to Venezuela.

With this rather inauspicious diplomatic beginning, President Nixon appointed a cabinet-level task force headed by Labor Secretary George Schultz. President Caldera was blunt in his report to the Schultz task force. The outcome would determine whether the United States considered Venezuela a "second-class friend."[33] Venezuelan Ambassador Julio Sosa-Rodríguez focused on Venezuela's 40-year record as a secure, friendly oil supplier and on Venezuela as a nation whose democratic faith was under constant challenge. The final report was disappointing to Venezuela. A special trip to the United States by President Caldera, which included a speech to a joint session of Congress, could only gain an increase of 35,000 barrels a day in the quota. The apparent reason for the denial of special preference was Venezuela's recent tax increases.[34]

In part, the official U.S. reaction was based on the twin assumptions that the world had an abundant and reasonably inexpensive supply of oil. Moreover, it was assumed that OPEC would remain an essentially ineffective organization. Both these assumptions were proven wrong in 1973. As late as 1972, mutual mistrust and suspicion still influenced Venezuela's relations with the transnationals and the United States. With reference to the transnationals, President Caldera reacted sharply when he felt that they were cutting production in Venezuela for no valid reason. In regard to a possible joint U.S.-Venezuelan development of what is now the Orinoco Heavy Oil Belt, disclosure in 1973 prevented President Caldera from initially proceeding with the project.

The year 1973 was an election year in Venezuela. The two principal populist parties agreed not to petroleumize the campaign by stirring up nationalist resentment that would have made rational decision making most difficult with regard to the then-anticipated nationalization of the industry. Both outgoing President Caldera and incoming President Pérez set the stage for nationalization by pointing out in an aggressive tone that the transnationals failed to incorporate new investment and maintenance in their plants and were attempting to maintain their production at minimal levels,[35] despite the Arab oil boycott of the West.

In early 1974, President Pérez installed a presidential commission on the reversion of the oil industry. In what was perhaps a trial run for the all-important oil industry, the Venezuelan Congress also authorized President Pérez to nationalize the iron-ore concessions by presidential decree. This was done, after months of negotiations, in December 1974. This nationalization of the U.S. Steel and Bethlehem steel investments in the Guayana region of Venezuela in 1975 was precedent-setting with regard to oil.

The Venezuela Guayana Corp. (CVG), created in 1960 and assigned to the Office of the Presidency of the Republic, was charged with the responsibility of planning and developing the resource-rich Guayana region of eastern Venezuela. The CVG has functioned since 1960, under the strong, technically oriented leadership of retired Gen. Rafael Alfonzo Ravard, as a holding company managing a number of diverse state and joint-ownership enterprises engaged in steel, aluminum, and electricity production. The nationalized iron industry was turned over to the CVG as its iron-mining subsidiary, *Ferrominera Orinoco S.A.* (Orinoco Iron Mining, Inc.). The CVG also had a long history of successfully forming mixed enterprises with Venezuelan and foreign capital. Perhaps of even greater significance, the CVG had earned the reputation of being a no-nonsense operation completely insulated from the normal populist practices of Venezuelan public enterprises.[36]

The road to nationalization between 1971 and 1976 was synonomous with "the Quest for Consensus" outlined by Martz. Others have indicated that by 1971, the transnational oil companies had hinted that they, too, would prefer nationalization to the potentially explosive rules of the game established by the 1971 Law of Reversion. Using the nationalization of iron mining as a model, what Venezuela intended to achieve was a moderate and nonradical nationalization of oil. Both Venezuela and the transnationals apparently agreed on two basic features of the 1976 global reality:

> Venezuela did not have an international marketing structure of its own that was sufficient to market 2.5 million barrels of crude oil a day.
>
> When Venezuela, given the depletion of its proven reserves, wanted to develop new reserves, it would need some legal technique for foreign assistance that would make future Venezuelan ventures attractive to foreign companies.[37]

What was being planned, therefore, was a "joyful nationalization," that is, one that would assure only winners and no losers.[38]

With the CVG model of an unchallenged, technically managed, politically insulated holding company, the only remaining issue was how the Venezuelan state-owned petroleum holding company would match the marketing and technology abilities of the transnationals.

In the 1973 elections, the two, serious, leftist, electoral coalitions—the "New Force" led by its presidential candidate, Jesús Angel Paz Galarraga, and the MAS Party whose presidential candidate was José Vicente Rangel—suggested that the entire oil industry be placed under the aegis of the CVP.[39] However, in the 1974–75 debate over the actual nationalization, no serious force suggested that the CVP could handle the absorption of 17 complex and distinct transnational operations and one small private firm. Mice may frighten elephants, but no one claims that they can devour them. The broad-based, multiparty presidential commission on nationalization, in the draft of Article 5 of the nationalization bill, adopted the precedent of the service contracts, stating that the state-owned holding company could "enter into necessary agreements" to contract for services. President Pérez changed this wording to read "to sign agreements of associations."[40]

The debate in the Venezuelan congress focused on the alternative wordings of Article 5. "Agreements of associations" implied that the transnationals, through operating agreements (*convenios operativos*), would retain influence in the operations of the Venezuelan industry and might actually return as part owners by means of mixed enterprises. Service contracts or agreements for technical assistance would not have this potential. The charge was made that all Venezuela truly would be nationalizing would be the risks of ownership and the heartaches of relating to its own workers while the transnationals were guaranteed the profits.

The government insisted that reality dictated the need for flexibility. The resultant contract accepted President Pérez's wording with the proviso that no partnership arrangement or any compromise of the principal of exclusive national ownership of the oil industry could occur without the approval of the Venezuelan Congress, meeting in a special, joint session. Former President Betancourt assured Venezuelans in 1975 that the possibility of "agreements of associations" would never herald a new period of "submissive surrender" of Venezuela's wealth to foreigners.[41]

NOTES

1. Stephen G. Rabe, *The Road to OPEC: United States Relations with Venezuela, 1919–1976* (Austin: University of Texas Press, 1982), p. 10.

2. Luis Vallenilla, *Oil: The Making of A New Economic Order—Venezuelan Oil and OPEC* (New York: McGraw-Hill, 1975), p. 30.

3. Edwin Lieuwen, *Petroleum in Venezuela: A History* (New York: Russell and Russell, 1959), p. 49.

4. Jonathan N. Leonard, *Men of Maracaibo* (New York: Putnam, 1933), pp. 84–85.

5. The Oil Ministry of Gumersindo Torres during the Gómez regime has been ignored by many Venezuelan and foreign scholars. None other than Rómulo Betancourt honored Torres for his "patriotism, honest and honorability" (p. 22). Betancourt wrote that when Torres died in 1946 and he was the president of the revolutionary junta, he stood vigil over the body of Torres "to the great surprise of all the mourners . . . most of whom were living remnants of the Gómez Regime" (p. 22). See Rómulo Betancourt, *Venezuela's Oil* (London: George Allen and Unwin, 1978), pp. 18–19, 22.

6. Ibid., p. 19.

7. John Lavin, *A Halo for Gómez* (New York: Pageant Press, 1954), pp. 416–17.

8. Rabe, *The Road to OPEC*, p. 44.

9. Ibid.

10. Bryce Wood, *The Making of the Good Neighbor Policy* (New York: Columbia University Press, 1961), pp. 260–76.

11. Ibid., p. 275.

12. Franklin Tugwell, *The Politics of Oil in Venezuela* (Stanford, Calif.: Stanford University Press, 1975), p. 44.

13. Ibid.

14. Wood, *The Making of the Good Neighbor Policy*, p. 263.

15. With regard to Gulf's Mene Grande problem, see Ibid., pp. 269, 279; to Exxon Creole's tax, Tugwell, *The Politics of Oil*, p. 46 note; and to the Gulf Oil agricultural mission, Ibid., p. 267.

16. Tugwell, *The Politics of Oil*, p. 48.

17. Kim Fuad, "Venezuela's Role in OPEC: Past, Present and Future," in *Contemporary Venezuela and Its Role in International Affairs*, ed. Robert D. Bond (New York: New York University Press, 1979), p. 125.

18. Rabe, *The Road to OPEC*, p. 158.

19. Fuad, "Venezuela's Role," p. 126.

20. Ibid., p. 134.

21. Ibid., p. 135.

22. Ruben Sader Pérez, *Explotación de Hidrocarboros por el Estado* (Caracas: Imprentz Nacional, 1966), p. 13.

23. Sader Pérez, *Explotación de Hidrocarboros*, p. 27, and Betancourt, *Venezuela's Oil*, pp. 72–73.

24. Tugwell, *The Politics of Oil*, p. 94.

25. Ibid.

26. Ibid., p. 77, and Rabe, *The Road to OPEC*, p. 163.

27. Betancourt, *Venezuela's Oil*, p. 53.

28. Ibid.

29. Rabe, *The Road to OPEC*, p. 163.

30. Tugwell, *The Politics of Oil*, p. 118.

31. Ibid., pp. 118–21.

32. Rabe, *The Road to OPEC*, p. 172.

33. Ibid., p. 173.

34. Tugwell, *The Politics of Oil*, p. 134.

35. Ibid., p. 123.

36. John R. Dinkelspiel, "Technology and Tradition: Regional and Urban Development in the Guayana" in *Venezuela 1969: Analysis of Progress*, ed. Philip B. Taylor, Jr., (Houston: Office of International Affairs, University of Houston, 1971), pp. 122–45.

37. James F. Petras, ed., *The Nationalization of Venezuelan Oil* (New York: Praeger, 1977), p. 117.

38. Stanford Rose, "Why the Multinational Tide is Ebbing," *Fortune*, August 1977, p. 118.

39. John D. Martz, "Policy-Making and the Quest for Consensus: Nationalizing Venezuelan Petroleum," *Journal of Inter-American Studies and World Affairs* 19 (November 1977): 498.

40. Ibid., pp. 492–93.

41. Betancourt, *Venezuela's Oil*, p. 45.

8

MANAGEMENT AND POLITICS: THE CASE OF OIL (1976–83)

POSTNATIONALIZATION DEMOCRATIC MANAGEMENT

In 1936, it was estimated that under the *caudillo* (warlord) Gómez, Venezuela received only about 7 percent of oil industry profits earned between 1920 and 1935. In 1976, the reserves and domestic installations of the industry became the property of Venezuela and a public enterprise. This turnabout, an affirmation of democracy's ability to slowly and hesitantly move a nation and to modify transnational forces, was the first oil nationalization since the Mexican example of 1938. Not all transnationals were happy with the negotiated compensation, nor were all Venezuelans pleased about the terms of the initial marketing and service contracts allowed as part of Article 5, yet it was a joyful nationalization: "The subsequent transfer of ownership and control went smoothly, receiving little worldwide publicity."[1]

Before approaching the issue of nationalization, the Venezuelans tried to make certain that if they were indeed "leaping into freedom," they wanted a reasonably soft cushion on which to land. The gradual "Venezuelanization" of the industry's work force meant that by 1976 well over 90 percent of the technical and managerial petroleum jobs were already held by Venezuelans. Article 5, which provided a safety line to the transnationals during the period of transition, was very much in the spirit of the post-1958 democratic experience; that is, cautious and deliberate incremental steps in all matters of great importance to the nation. There were no grand transformations or grand leaps into freedom from 1958 to 1976, nor were any intended with regard to Venezuela's oil industry.

Evaluating the performance of public enterprises such as Petroleum of Venezuela S.A. (PDVSA) is not an easy task.* "There is no generally accepted formula with which to do the evaluation."[2] Venezuela's oil people have scrutinized the experience of nationalized oil industries in Mexico, Argentina, and Brazil and believe they have come up with the model best suited to Venezuela's unique needs and conditions. Venezuela's petroleum nationalization, unlike Mexico's, was not sudden, allowing PDVSA's future management to study the worldwide experience with public enterprises over the previous several years.

The key dilemma of public enterprises revolves around two questions. First, what should the role of the profit motive be in the management of such enterprises? According to Brazilian Minister of Industry and Commerce João Camillo Penna, profit can no longer be considered a dirty word when the government's business is managing big business.[3] However, profit cannot be the sole criterion for judging public enterprises. Public enterprises have critical social and developmental functions to assure society's proper operations rather than their own profitability, and these externalities must be considered. Price setting of public enterprises can, for example, be an instrument for the redistribution of national income, and its prices may be set low in order to stimulate other economic activities such as a petrochemical industry. In addition, its personnel policy can legitimately be considerable overemployment, not only to meet short-term needs of lowering unemployment but also to serve as a training ground to upgrade the skills of the national labor force. Finally, public enterprises can function as self-contained models of social and organizational experimentation. Ideas such as worker participation in management, innovative community design, or educational programs might best be tried in such controlled environments before being unleashed on society.[4]

---

*Petróleo de Venezuela once used the acronym PETROVEN. However, when this state-owned company was being registered under the Commercial Code, the existence of a small private company with a similar name came to light. For this reason, the initials PDVSA (pronounced PeDeVeSA) were adopted.

The second question is whether the operations of a key public enterprise such as PDVSA can proceed as an island of rationality in the sea of praetorianism, populism, and traditionalism that is the reality of Venezuela. The notion of an "island of rationality" is derived from the work of the World Bank in the 1950s and 1960s. Because of its unfortunate association with the "enclave" of the transnational days and its vulnerability to the accusation of being an instrument of "dependency," it has been replaced by the more neutral "self-contained-unit" concept. Rephrased, the question is how autonomous the public enterprise should be. In discussing the evolution of democratic management of the postnationalized oil industry, there are two facets to this concern: the relationship of PDVSA as a public enterprise to the elected political authority, and how the public enterprise and the political authorities relate to Venezuelan society.

With regard to the profit motive, the oil industry in Venezuela has accounted for 70 percent of government revenue, 26 percent of gross national product, and 90 percent of the nation's foreign earnings. Respect for the profit motive is one overwhelming concern of the management of the industry. Venezuela has had little margin for error in its "muddling through" regarding the nationalized oil industry and has had no opportunity for populist "mucking around". Since the public treasury relies heavily on PDVSA, there is little prospect that PDVSA could go to it for emergency bailing out. In the words of the chairman of the Senate Committee on Administration and Services, Social Christian Party (COPEI), Sen. Valmore Acevedo, "Democratic oversight of the PDVSA must, above all, not kill the goose that lays the golden eggs. It will however, insist that the goose be *criolla* [Venezuelan]."[5]

The accuracy of Figure 8.1 would not be accepted by every key Venezuelan. One top official of the Ministry of Energy and Mines in a 1980 interview with the author insisted that the consultative National Energy Council has a central role to play in the democratic oversight of the industry.

In the figure, the principal, central line of political policy making for the oil industry runs from the elected president through the Economic Cabinet (the ministers of Finance, Planning, Development, and Energy) to the minister of Energy. The minister of Energy acts as president of PDVSA's Stockholders Assembly.

Congress moved to establish its oversight responsibilities over the nationalized industry in public hearings called in response to the

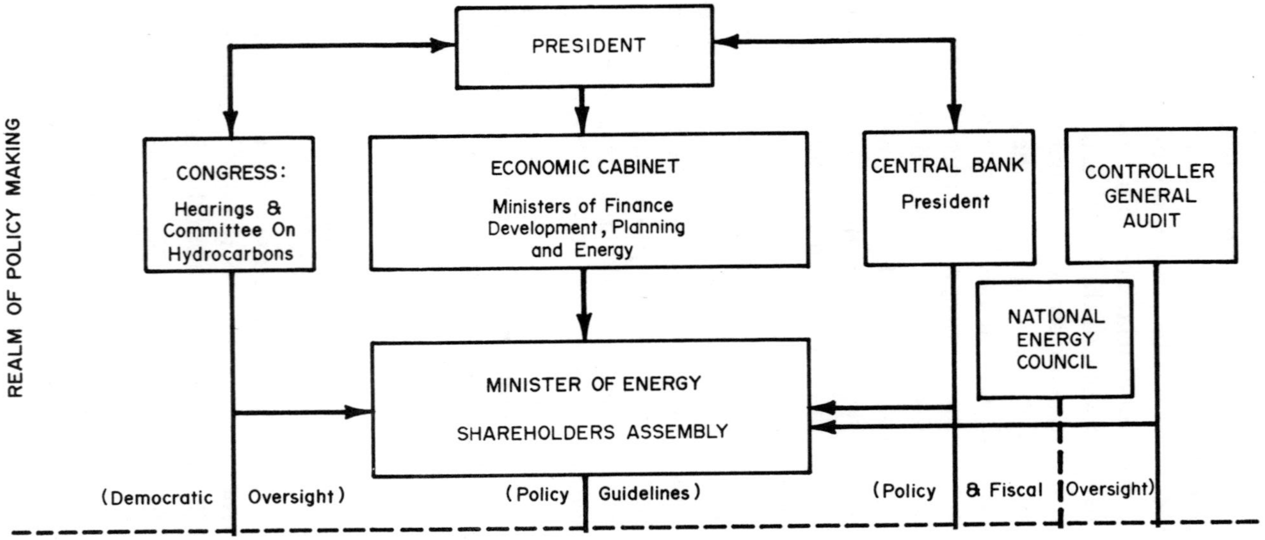
REALM OF POLICY MAKING
PRESIDENT
CONGRESS:
Hearings & Committee On Hydrocarbons
ECONOMIC CABINET
Ministers of Finance Development, Planning and Energy
CENTRAL BANK
President
CONTROLLER GENERAL AUDIT
NATIONAL ENERGY COUNCIL
MINISTER OF ENERGY
SHAREHOLDERS ASSEMBLY
(Democratic Oversight)
(Policy Guidelines)
(Policy & Fiscal Oversight)

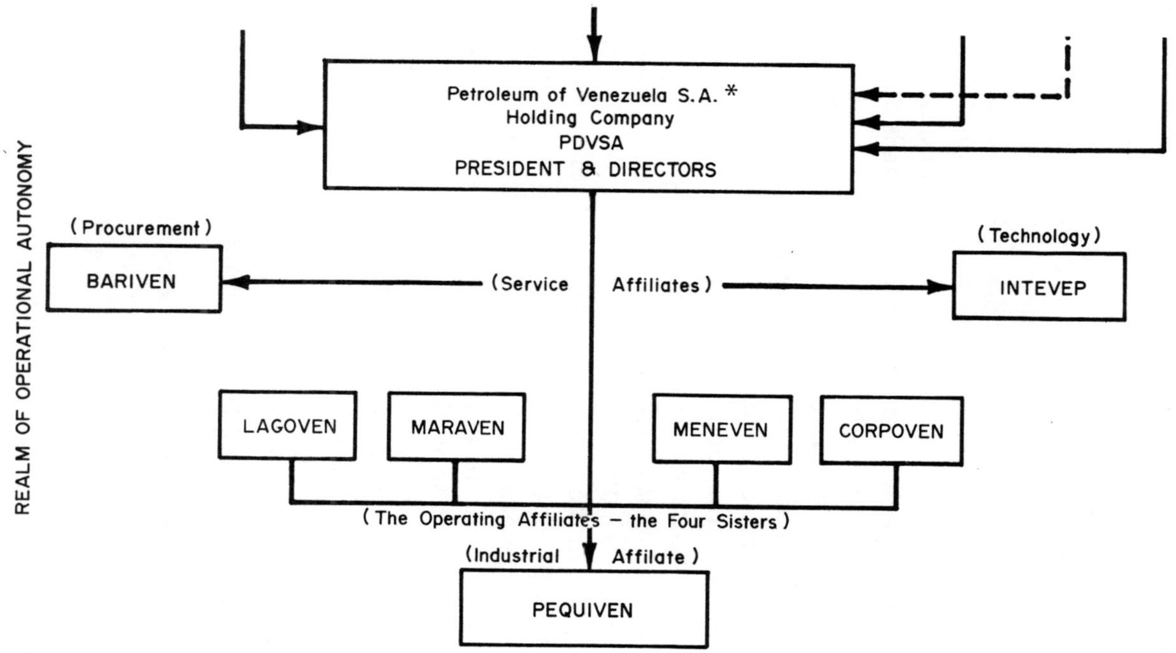

Figure 8.1. Schematic Chart of Democratic Management of Oil

*The international affiliates—PDVSA-USA (New York), PDVSA-UK (London), and BARIVEN-USA (Houston)—are not included in this schematic chart. There are also PETROLATIN, created in 1982 by PDVSA, Mexico's PEMEX, and Brazil's PETROBRAS.

first annual report submitted by PDVSA in early 1977. These hearings highlighted what has been a constant and perhaps institutionally based strain between the elected legislature and public enterprise, namely PDVSA's use of secrecy. The maintenance of corporate and technological secrets is standard practice in private business. PDVSA's managers, while denying impropriety in their dealings, have continually refused to reveal publicly the contents of many of their business arrangements, including the technologies and sales contracts they have signed with the transnationals. While the parties on the left sharply criticized PDVSA secrecy, the two principal populist parties showed great restraint in publicly pressing PDVSA for details during the years following nationalization (1977–80).[6]

The dictum of not killing the goose has tended to be amplified to mean "don't unnecessarily molest it during its awkward childhood." For one thing, congress failed to adequately upgrade its research and staff capabilities, preventing that body from exercising proper oversight. There is nothing in Venezuela comparable to the U.S. Government Office of Technology Assessment, Library of Congress, which has so ably served the U.S. Congress.

The current president of the Central Bank (BCV), Leopoldo Díaz Bruzual, has also assumed a critical role in the fiscal and policy oversight of PDVSA. Díaz Bruzual assumed the BCV presidency in April 1981 after he publicly attacked the populist economic policies advocated by the Economic Cabinet as president of the Venezuelan Investment Fund (FIV). Díaz Bruzual's polemical style has made him a folk hero in the eyes of some and President Hererra's economic-affairs "storm trooper" in the eyes of others. Many in the business community are disturbed by his activist and personalist style of discussing economic issues, preferring a more serene Central Bank president.[7] Many accept that his actions are a reflection of the president's concerns.

Venezuela has a Controller General of the Republic charged with carrying out external audits of all public enterprises with the intent of preventing fraud. When fraud or mismanagement is suspected, the Controller General presents his concerns to both the elected Congress and the general public. While these published audits and investigations invariably criticize other state enterprises for nonchalant bookkeeping, unexplained losses, and frequent scandals, the management of the PDVSA has been untainted by any sustained charge or hint of scandal during the critical years of the petrobonanza and the more recent austerity (1976–83).

Given the constant political controversy that has swirled around the PDVSA and the immense funds it autonomously administered, this scandal-free record is a most remarkable achievement. In light of the scandal that wracked Mexico's PEMEX in 1983 and the utterly dismal performance of the other Venezuelan state enterprises, this quality of both the top managers of the PDVSA and of the system of democratic oversight deserves an honorable mention in this volume.

The role of the consultative National Energy Council is problematic. Although as a collegial body it has never formally met, it has become the principal forum through which university intellectuals and the radical-protest political parties, which are relatively weak in congress, articulate their concerns. If the activism of the Central Bank is dependent on the president, that of the National Energy Council is dependent on the nation's press. That a Communist (PVC) leader such as Radames Larrazábal, Socialist (MAS) academic Francisco Miéres, People's Electoral Movement (MEP) Alvaro Silva Calderón, and independent oil expert Aníbal Martinez are members of the National Energy Council give their pronouncements increased legitimacy. The electoral weakness of their political parties and the decision of their party leaders not to place their petroleum experts in congress has necessitated use of the National Energy Council. Despite PDVSA management's insistence that it does not engage in political polemics, it has responded to the technical questions put forth by members of the council.

One of the constant ambiguities of the politics of postnationalization has been the relationship between the minister of Energy and the president of PDVSA. To put it mildly, some of PDVSA's directors and top managers believe that there is no need for the ministry in its present form, while many in the ministry still hint that the adoption of the Guayana Corp. (CVG) holding-company model for oil was not necessary. President Pérez's minister of Energy, Valentine Hernández Acosta, was unhappy with the decision to grant PDVSA the planning, coordinating, and supervisory authority for the industry rather than the ministry. He was vocal in expressing his unhappiness over the choice of organizational model. The more responsibilities PDVSA acquires, the less remains in the hands of the ministry. Minister Hernández Acosta was quoted as stating in 1976:

> We have to be careful with Petroleos de Venezuela [*Petroven*, now PDVSA] with respect to the technical matters and the development of the industry as we have been with the concessionary companies. . . . The experience that we have had is that the Ministry had more troubles . . . with the CVP than with the concessionary companies. . . . We cannot let PDVSA treat the subject from the purely commercial point of view.[8]

Apparently because of the strong, no-nonsense, technically oriented role that Gen. Alfonzo Ravard assumed during his presidency of the CVG, many political leaders opposed his nomination as president of PDVSA. Julio Sosa Rodríguez, who had been President Caldera's ambassador to Washington and who had led Venezuela's unsuccessful fight to be included in the hemispheric system of preference, was one of the nominees preferred by the political parties. Alfonzo Ravard's international stature, his national reputation for technical competence, and the overwhelming support he had among the private-business sector apparently secured his nomination. Upon assuming the position in 1975, he demanded and received a maximum degree of operational autonomy for the industry and a strong presidential commitment to keep political hands off its operations.

According to PDVSA President Rafael Alfonzo Ravard, maximum corporate autonomy was granted PDVSA. While 100 percent state-owned, PDVSA "has been structured and operates as a commercial company. . . which seeks to obtain maximum economic benefits for its sole stockholder . . . the Venezuelan State."[9] PDVSA is registered under the Commercial Code of the nation and is subjected to the enforcement of strict accounting standards. PDVSA operating affiliates are also subjected to the same taxes and tax-accounting procedures that the private foreign concessionaires were. The second aspect of the PDVSA operational autonomy has been the nation's remittance to it of 10 percent of the net income from export sales for the development of the oil industry. The Venezuelan government in the 1976–80 period allowed PDVSA to retain just about all its net income after paying taxes rather than having dividends declared to the nation. This has enabled PDVSA to build up a monetary reserve or autonomous investment fund estimated to be in the \$8 billion–\$9 billion range. In late 1982, the PDVSA lost control over this fund.

The immediate problem facing PDVSA in 1976 was the organization of the 14 operating companies into a rational productive entity. Fourteen operating companies were too many to manage, but given the different histories and corporate identities of the once-competitive concessionaires, it would have been most difficult to have only one company. In a two-step process of consolidation, the 14 companies were reduced to "Four Sisters" in 1978.

The retention of four operating affiliates and the maintenance of a healthy sense of competition among them has been a key element in PDVSA's control of its operating affiliates. One of the key management and control problems of public enterprises has been their monopoly status, which often has created "an oil industry too powerful for the nation's political authorities to control and which, therefore, might come to control them."[10]

The maintenance of the Four Sisters has not only facilitated efficient operations by means of competition but also has secured democratic control. At the end of 1978, PDVSA President Alfonzo Ravard announced the culmination of the nationalized industry's juridical consolidation. Alfonzo Ravard's public reason for consolidation was, as always, a technical one, "to optimize the use of the infrastructure and the productive installation" of the industry.[11] The Four Sisters:

> The big American sister, LAGOVEN, was essentially Exxon's *Creole* subsidiary. LAGOVEN accounts for 40 percent of national production, Creole accounted for 20 percent of Exxon's world production in 1976. LAGOVEN President Guillermo Rodríguez had been with Creole since 1945. LAGOVEN had been tightly controlled by Exxon and simply had to change its loyalty from New York to Caracas.
>
> The European sister, MARAVEN, was essentially Shell of Venezuela. Shell allowed considerable discretion to its local management, encouraging it to be "creative and innovative." MARAVEN President Alberto Quiros Corradi projects an autonomous and intellectual image. He wrote a provocative article, "Energy and the Exercise of Power," for *Foreign Affairs* in 1979. Oil executives have traditionally shunned politics, while Quiros Corradi has openly written about the subject.
>
> MENEVEN has been called the go-getter sister. Its 1979–81 management under Juan Chacin faced the organizational dilemma of blending four companies into one, although the bulk of what is

> now MENEVEN was Gulf's *Mene Grande* subsidiary. Chacin spent his early career in the formerly state-owned CVP and CORPOVEN before moving over to head MENEVEN. He had to make hard decisions while with CVP, such as trimming its payroll by 30 percent.
>
> The fourth and newest sister, CORPOVEN, has been described as a "potpourri of several companies in search of a corporate identity." Its president, Frank Alcock, spent his early career with Mobil Oil's small Venezuelan venture. The former CVP is the largest component that went into making CORPOVEN.[12]

To demonstrate the ambiguity that separated the realm of political policy making from the realm of corporate operational autonomy, in 1978, the year in which PDVSA completed the consolidation of its operating affiliates, two new affiliates joined the holding company. The state enterprises that had been charged with developing the nation's petrochemical industry had become "infamous for poor performance" for more than a decade.[13] In 1976, Alfonzo Ravard and PDVSA resisted the affiliation of this state enterprise because of the concern that the corruption and inefficiency associated with petrochemicals could be catching. Gen. Alfonzo Ravard announced in 1978 that the government also assigned to PDVSA the state petrochemical company, PEQUIVEN.

PEQUIVEN, reorganized as the industrial affiliate of PDVSA, has had a logical basis. The ultimate Venezuelan intent is to convert raw materials, such as oil, into manufactured goods, employing Venezuelan talent and know-how in the process, before selling the end product on the world market. Even if Venezuela earned $50 per barrel for its oil in 1990, it would prefer to convert that barrel of oil into fertilizers, paints, and synthetic plastics before marketing it. Rather than succumbing to the PEQUIVEN virus of corruption and mismanagement, between 1978 and 1981, PDVSA apparently imposed its virtues on this affiliate. PEQUIVEN showed a profit in 1982 despite the softening petrochemical market.

The second affiliate was a more welcomed addition. The Venezuelan Petroleum Technology Institute (INTEVEP) became a service affiliate of PDVSA, although it was not a traditional enterprise selling a clearly defined product. The affiliation of INTEVEP in 1977 may have been in part a defensive move to demonstrate that the PDVSA rather than the Ministry of Energy could handle complex technological studies such as that involved in developing the Orinoco

Heavy Oil Belt. INTEVEP has also become involved in representing PDVSA in environmental questions. The mission of INTEVEP is applied research. In a very un-PDVSA manner, INTEVEP occasionally suggests that its primary function is political, that is, freeing Venezuela from technological dependency.

The creation in 1980 of a second service affiliate, BARIVEN, an exterior-materials and equipment-procurement corporation, was partially undertaken for similar defensive reasons. BARIVEN, which exists as a U.S. company, BARIVEN U.S.A., headquartered in Houston, Texas, is to attempt to rationalize and standardize the materials and equipment purchases of the four operating affiliates. Its creation is a clear indication that the relationship between PDVSA and the Four Sisters has at times been rather stressful. In rationalizing the procurement of materials from abroad, the next logical step would be to consider their substitution by national production. BARIVEN should shortly play a critical role in relating PDVSA's material input needs to CVG's industrial output.

The question of the profit motive regarding PDVSA is involved with the broader question of marketing Venezuelan oil. The fear that the giant transnationals, Exxon and Shell, would control PDVSA by controlling its international marketing and distribution engendered as much controversy in the oil-nationalization debate as did the concern over the service-contract association clause, Article 5. Apparently there was a demand to create a state-owned marketing company in 1976, but this has been rejected. Although PDVSA has an International Marketing director and initially used the traditional marketing ties between nationalized company and the former transnational owner (LAGOVEN-Exxon, MARAVEN-Shell), international marketing occurs within the framework of a broad political as well as commercial strategy.

Minister of Energy Calderón Berti called the early 1980s' marketing strategy one of "globality." Venezuela under globality wants to build on its reputation of being a reliable and honest supplier who, with the credibility the Orinco Heavy Oil Belt assures, can offer believable five- to ten-year supply commitments. Venezuela also wants more than dollars for its oil; it wants to develop secure long-term partnerships with industrial nations in which Venezuelan oil can be exchanged for technology and technical assistance. For example, "discussion with Japan has sought technical aid in planning Venezuelan mass transit systems in exchange for guaranteed oil supplies."[14] In addition to seeking innovative mechanisms for marketing its oil and thus bypassing the transnationals, the political strategy includes securing new markets.

Some long and hard negotiations have occurred with France, Italy, Brazil, Spain, and the Federal Republic of Germany over the possibility of jointly financing refining capability in Europe to handle Venezuela's heavy oil. These complex negotiations have been on a government-to-government level as well as between PDVSA and state-owned European oil companies. The 1983 signing of an accord between PDVSA and the German public-owned VEBA Petrochemical Corp. has been called the most significant accord signed by Venezuela with a petrochemical company not previously represented in Venezuela. The accord enables Venezuela to sell 100,000 barrels of oil per day to Germany. The PDVSA now is a partner with VEBA in owning two oil refineries in the Federal Republic of Germany. Before the PDVSA and VEBA signed their accord, five years of joint research on Venezuelan heavy oil was undertaken by both publicly owned companies.[15] In 1981, the PDVSA signed an accord with the French, public-owned, petroleum corporation, however, this arrangement has not had the same success as the PDVSA-VEBA accord. The PDVSA, as a public enterprise, does not involve itself in government-to-government accords.

The complex government-to-government oil diplomacy is legitimately in the hands of the Ministry of Energy and the Economic Cabinet, which includes the ministers of Energy and Technology. For example, the 1981 Brazilian-Venezuelan trade negotiations involved parallel-bartering arrangements of Brazilian sugar and Venezuelan oil. There was also mention of joint financing involving the conversion of one of Brazil's largest refineries to handle Venezuela's heavy oil. Brazil's purchases of Venezuelan oil more than tripled in the two-period 1980–81, reaching about 100,000 barrels per day in 1981. This was partially due to the Persian Gulf War. Since the Organization of Petroleum Exporting Countries (OPEC) frowns upon bartering of oil as a means of giving price discounts, parallel bartering does not directly involve the discount bartering of oil.

In the Western Hemisphere market, Venezuela has been a major proponent of bilateral and multilateral government-to-government efforts at creating hemispheric energy cooperation. In a speech delivered in New York in March 1980, Minister Calderón Berti explicitly mentioned that in light of the increasingly delicate international political situation, "hemispheric [energy] security" concerns all in the region.[16]

Venezuela was a prime mover in the creation of the Latin-American Energy Organization (OLADE) and the Inter-American Energy Development Fund, which seeks to create broad hemispheric commitments and capabilities to resolve the region's energy deficiencies. Venezuela's hemispheric energy diplomacy has adopted the strategy of riding two different horses at the same time. One horse is that of developing a common petroleum policy with Latin America's other large exporter, Mexico, while the other is securing its U.S. oil connection. The two horses have not always been heading in the same direction nor moving at the same speed.

In August 1980, Mexico and Venezuela signed an energy agreement to supply nine energy-poor and soft-currency nations with up to 160,000 barrels per day. Both Mexico and Venezuela agreed to finance its share of the venture through joint low-interest loans. In 1981, this binational energy agreement was extended to include Haiti and Belize. The energy-cooperation program had mutual security purposes as well as economic ones; that is, calming the turbulent waters of the Caribbean basin. Cuba has been receiving Venezuelan oil, according to a separate arrangement, on a month-to-month basis.

Venezuelan relations with the United States continue to be affected by the U.S. failure to include Venezuela in the hemispheric-preference system it maintained during the 1950s, a period of world oil overproduction. Venezuela is also concerned that Mexico, by necessity, will compete with it for a larger share of the U.S. market even if Middle East production remains disrupted by the instability of that region. Such politicians as Julio Sosa Rodríquez, who has been personally involved in Venezuela-U.S. relations, have become strong advocates of Venezuela assuming a cautious approach to the U.S. market. Sosa Rodríquez may have suggested that the United States take a more long-term perspective toward the Orinoco Heavy Oil Belt.[17]

In the 1981–82 period of fiscal austerity, concerns about the long-term marketing of oil and the long-term financing of future oil development came together. PDVSA management and Democratic Action (AD) oil spokesperson, Deputy Celestino Armas, have advocated utilizing Article 5 to get partial U.S. funding for the risky and costly up-front investments necessary for the development of the Orinoco Belt. The Armas-AD proposal would seek either private transnational or public cofunding.[18] Under the Carter administration, serious thought was given to having the newly created U.S. public enterprise, Synthetic Fuels Corp. (SFC), fund part of the Orinoco development from its $2 billion fund created as part of the U.S. Energy Security Act.

Former U.S. Energy Secretary Charles W. Duncan visited Caracas in 1980. Several scientific and technological cooperation agreements were signed in the 1979–80 period, dealing with energy, agriculture, health, and scientific exchanges. A number of joint Venezuelan–U.S. projects have been proceeding under these arrangements. The Reagan administration has de-emphasized government-to-government economic relations and has hinted at terminating both the SFC and the Department of Energy. This lack of continuity in U.S. policy has assisted those in Venezuela who advocate diversifying its markets and de-emphasizing the United States. While the Venezuelan-German negotiations to trade technology for oil have reached fruition, those between Venezuela and the United States have "fizzled," according to one key Venezuelan source.

By the end of 1981, 50 percent of the 1.4 million barrels per day Venezuela exported was sold to new clients. Japan emerged as a new customer as part of its efforts to diversify its sources of oil.[19] The acquisition of other new customers, such as the Federal Republic of Germany and Brazil, and the use of innovative techniques of selling its oil have demonstrated that the policy of government-to-government oil diplomacy performs well. The marketing of oil in 1981, a period of oil glut, was not easy, but the basic soundness of the Venezuelan policy of keeping all global options open proved a good one in this period of fiscal austerity and deflated aspirations.

PDVSA's childhood has been marked by its meeting three great challenges: converting its oil refining and marketing operation from one based on light oil to one based on heavy oil, developing off-shore exploration and production capabilities as one option to find new reserves, and going full throttle to develop the extraordinarily heavy oil that abounds in the Orinoco Heavy Oil Belt. A fourth, generally unpublicized, challenge has been the rescue and basic redevelopment of the Lake Maracaibo oil fields. Basically, to rescue Lake Maracaibo production means to use new technology (tertiary recovery), which should prolong the life of these old fields. The dangers involved in the new technology and the possibility that the lake may flood communities such as Lagunillas have brought on the possibility of a basic redesign of the urban settlements that grew up along the east coast of Lake Maracaibo.

During this period of a most-challenging childhood, PDVSA has also received relatively high marks in overcoming the four principal problems that have plagued public enterprises worldwide: politicalization, labor problems, the loss of competent technical and managerial personnel, and a decline in corporate efficiency.[20]

In regard to labor and technical personnel, it is noteworthy that peace has been partially secured by having two, senior labor leaders with long experience in the Federation of Petroleum Workers of Venezuela, Manuel Peñalver and Raúl Henriquez, placed on PDVSA's board of directors. They serve as representatives of the labor force and apparently do not play an active role in any of the company's permanent committees. Perhaps of critical importance has been the fact that petroleum-sector wages, fringe benefits, and pension plans are the best in the nation. The cost of the industry's labor peace has not been insignificant.

There is considerable resentment in communities located near PDVSA operations over the privileges granted oil workers as part of recent collective-bargaining agreements. The author was told in 1980 that some staple food products in oil company commissaries were available at 1940 prices. The collective-bargaining agreements have led to the installation of private telephones in oil-worker homes, while professionals living in adjacent communities are required to pay considerable bribes to get telephones installed in their offices. The fences and guards around the formerly foreign-owned oil camps are still there, even though the communities are now called "residential units." The schools, hospitals, and recreational facilities available in these residential units are far superior to those available to the general public.

Meeting the previously mentioned challenges and perhaps responding in part to pressure for jobs, employment in the industry grew by more than 17,000 between 1976 and 1981. Industry personnel in 1981 exceeded 42,000 compared to the 24,000 estimated in 1975. Industry employment in 1981 alone grew by 4,500 jobs.

The weakened world-oil market and the third year of economic recession set the stage for the politicalization of the management of oil. The taboo against involving PDVSA management in political debates and the gentlemen's agreement between AD and COPEI not to petroleumize their partisan disputes ended in 1981. In a public address, the AD president and party presidential candidate in 1968, Gonzálo Barrios, severely criticized the management of PDVSA for its "ultraextravagant" investment program and the life styles of its senior executives. The AD chair of the Senate Committee on Hydrocarbons and a former Interior minister, Octavio Lepage, also publicly released the contents of two questionnaires sent to Minister Calderón Berti and PDVSA President Alfonzo Ravard.[21]

Perhaps adding insult to injury, Sen. Lepage indicated that he would accept hearings behind closed doors in order to resolve the matters brought up in his questionnaires. In addition to asking for an accounting of overall operating costs for PDVSA and the detailed evaluations of the program to develop the Orinoco Heavy-Oil Belt, Lepage requested information on the salaries and bonuses paid to top executives, consultants, and to personnel in PDVSA's foreign missions. A detailed analysis of the increase of PDVSA's personnel since 1976 was included in the questionnaires. COPEI Secretary Gen. Eduardo Fernández, speaking in the name of the party, appealed for all to maintain petroleum matters separate from partisan conflict and not to disturb the continuity of the industry.[22] Some have noted that those in COPEI identified as *calderistas* did not defend PDVSA and the Hererra administration's oil policy.

The heated congressional debate was only one component of a broader national debate that has been swirling around PDVSA since the decision to develop the Orinoco Heavy-Oil Belt. By 1980, the internal Venezuelan debate over the Orinoco Belt attracted international attention. For example, a lengthy article on the Orinoco development appeared in The Washington *Post*. The *Post* article contained interviews with both oil Minister Calderón Berti and respected oil geologist Anibal Martinez, one of the opponents of rushing in to develop the Orinoco.[23]

The managerial impulse has always preferred quiet, delimited, technical-based dialogues. The highly emotional issues involved in the development of the Orinoco Heavy-Oil Belt and the sense that the economic future of the nation was at stake simply did not allow the issue to be contained in the genteel world of private discussions and planning dialogues. In a real sense, the accountability for the proper development of the Orinoco was owed to the nation, not just to the political authority of the day. Illustrative of the emotional content of the debate was the July 1980 speech of Minister Calderón Berti to a meeting of the industrialists who produce parts for the oil industry: "We cannot allow a repetition of Cabimas and Lagunillas where decades of oil production contributed little or nothing to the surrounding region."[24]

The debate in the public arena and the issues presented overwhelmed all attempts to limit the discussion to technical feasibility and cost effectiveness of the project. It was not just an administrative question of privately arm twisting PDVSA to become sensitive to social concerns. The public debate increasingly focused on allegations of secret deals with U.S.-owned transnationals and of a U.S. inclusion of the Orinoco Valley in the "strategic reserves of the United States." In reference to the publication of a staff report to the U.S. Senate Committee on Energy, *The Geopolitics of Oil*,[25] and the election of Ronald Reagan, Venezuelans were told to prepare for the worst.

The secret agreement to develop the Orinoco was allegedly made with the Bechtel Engineering Group. By early 1981, Bechtel's counsel, Caspar Weinberger, was the U.S. defense secretary. Two former U.S. Central Intelligence Agency directors, Richard Helms and John McCone, had both worked for Bechtel. Bechtel also had had close ties to Exxon.[26] Finally, Bechtel President George Schultz had been Nixon's Labor secretary and had headed the 1971–72 task force on oil imports that had disappointed Venezuela. (Schultz left Bechtel in July 1982.) Cartoons in Venezuela featured crocodile versions of Uncle Sam emerging from the Orinoco and of Caspar Weinberger flying overhead in an F-16 seeking to seize the Heavy Oil Belt. PDVSA representatives have admitted that the use of the term "megaproject" to describe LAGOVEN's program for the Orinoco stirred unnecessary fears because LAGOVEN was formerly Exxon's Creole.

The response of PDVSA representatives to these issues was to question what all this had to do with the fact that Bechtel was the world's most experienced engineering firm with regard to developing nonconventional heavy oils. These PDVSA sources indicated that the radical regime in Algeria was quite satisfied with Bechtel and that Bechtel had already been at work on Canada's Atabasca tar sands. Given the emotions of the moment and the fact that the AD majority faction in the College of Engineers and the recession-ravaged business community were insisting that PDVSA hire Venezuelan and buy Venezuelan, Bechtel did not get the contract. A three-year engineering contract was given to a second U.S. company, the Lummus Group of New Jersey, which allegedly has ties to one of Venezuela's large family-owned business groups.

In setting the guidelines for the industry, Minister Calderón Berti established the twin goals of globality and the deepening of the meaning of nationalization by strengthening ties between the industry and the nation. Oil was to be an instrument in Venezuela's comprehensive development, not just a source of revenue. In order to demonstrate that PDVSA was indeed accountable to the nation, Alfonzo Ravard announced that the Orinoco Valley was to be developed in an economically balanced and socially integrated manner and that he would not allow fenced-in oil camps. PDVSA signed an agreement with the Ministry of Environment and Natural Resources to establish a framework within which to develop the Orinoco back in 1980. This was followed in 1981 by efforts to establish a working commission with the CVG to facilitate purchases from that public, regional-development agency's industrial affiliates. In 1983, given the world oil market, Venezuela officially announced that it was delaying, or "redimensioning," the Orinoco heavy-oil development.

PDVSA had also begun the rescue and redevelopment of the Lake Maracaibo oil fields. Unlike the situation in the Orinoco, the industrial infrastructure was in place in the Lake Maracaibo area. However, MARAVEN, PDVSA's affiliate in charge in that region, would have to help finance basic urban redevelopment.[27] The deepening of the meaning of nationalization has meant that MARAVEN has acted as part of a mixed public-agency effort for the redevelopment of the east coast of Lake Maracaibo, not as an isolated public enterprise indifferent to the fate of local communities. The replanning of the Lake Maracaibo region has involved the elected municipal councils and the grass-roots business associations and labor-union federations. This redevelopment has been exemplary as a model of democratic planning. The June 1983 decision to "redimension" the Orinoco development meant that Lake Maracaibo would remain the center of oil production. The reality of democratic politics seems to indicate that the community needs of the people of the east coast of the lake can no longer be ignored.

The August 1981 PDVSA Shareholders Assembly gave the minister an opportunity to announce an additional aspect of the deepening of the ties between industry and nation: the operating affiliates of

PDVSA would have their headquarters decentralized out of Caracas. The decentralization of administration was long a promise of Venezuela's politicians, and the PDVSA was to be regionalized; MARAVEN was to relocate in the eastern city of Puerto La Cruz, while Pequiven would go to Maracaibo. The fact that MENEVEN's Vice-president Gustavo Coronel first read about his being regionalized in the morning newspaper was the final cause for his resignation.[28] The examples of MARAVEN's institutionalizing a good working relationship with the communities of the east coast of Lake Maracaibo, MENEVEN's harmonious move to Puerto La Cruz, and LAGOVEN's sensitivity to the agricultural interests of the Orinoco Belt demonstrate PDVSA's ability to adapt to its regionalization.

The Shareholders Assembly also restructured the board of directors of PDVSA in a way interpreted as a slap to Gen. Alfonzo Ravard. Men considered personally loyal to Minister Calderón Berti replaced men with long careers in the industry. One dismissal even led to a public protest. Alfonzo Ravard was retained as president under a two-year contract. The new board member in charge of purchasing, Humberto Peñaloza, indicated that buying Venezuelan would be emphasized. Rumors that Gen. Alfonzo Ravard would shortly resign had begun to make their rounds.

The 1981 end-of-year report of the BCV contained further criticism of Alfonzo Ravard's management. According to the BCV's Díaz Bruzual, Alfonzo Ravard had produced more new employees than new wells, and PDVSA was an inefficient and overpaid organization. According to the BCV, PDVSA workers received three times the salary paid normal civil servants.[29] Alfonzo Ravard answered Díaz Bruzual's criticism "with serenity," pointing out that PDVSA workers were not "civil servants" but were covered by the wage traditions of the oil industry and the 1975 Nationalization Law, which protected the lucrative salaries of the workers through collective bargaining.[30] The Díaz Bruzual-Ravard dialogue expanded in 1981 and 1982 into a national debate over the future of oil.

In 1981, President Caldera's Energy minister, Hugo Pérez La Salvia, attacked the PDVSA holding-company concept. Contrary to the PDVSA belief of preserving a sense of healthy competition among the operating affiliates, Pérez La Salvia suggested that the Mexican and Brazilian experience with only one state enterprise should be considered.[31]

The basic controversy involved with deepening the meaning of the 1976 oil nationalization in 1981 appeared to be over control of PDVSA international reserves, estimated between $8 billion and $9 billion. Perhaps in preparation for the 1983 elections as well as in response to the increased difficulty Venezuela faced in servicing its international debt, Central Office of Coordination and Planning director, former Planning Minister Ricardo Martinez suggested that some of this enormous reserve be used for necessary development projects not directly related to petroleum. Martinez reasoned that with PDVSA's excellent record, the enterprise should have no problem in securing external financing for its projects. Martinez allegedly argued, "It does not make sense—that country roads should be built with London credit when the oil industry could get loans at much more attractive rates" than the nation.[32]

In early 1982, PDVSA response to the unprecedented challenges presented to its autonomy (according to MARAVEN President Alberto Quiros Corradi, "The honeymoon was over") was more assertive exercise of its policy of only dealing with technical concerns. This policy has had an international as well as a domestic dimension. Given the prospects for a more competitive price situation in energy in the near future, PDVSA worked to assure the world market that heavy crudes and tar sands are less costly to produce than deep off-shore oil in the North Sea or off the tragic-comic Falklands-Malvinas. It should also have a clear price advantage over the conversion of coal into gas.

PDVSA arranged to have the U.N. Institute for Training and Research cosponsor the second Caracas Conference, in February 1982, on heavy crude and tar sands. The papers presented at the conference were very supportive of the PDVSA contention that the Orinoco heavy oil was indeed the "gooey gold" of the future. Venezuela was said to have 55 percent of the recoverable proven reserves of heavy oil.

According to PDVSA sources, the cost of the high energy consumption needed to inject steam to extract the heavy oil is the only remaining obstacle to a commercially successful development. However, even with today's technology, Orinoco crude has been considerably less costly to produce than synthetic coal conversion, deep off-shore oil, and Rocky Mountain tar shale.[33] Some in the Ministry of Energy and the Ministry of Environment have advocated solar energy as a means of lowering energy costs in this sun-soaked region.

Others have indicated that the residue sludge left from extracting heavy oil could also be used to fire specially designed steam-generation plants, avoiding the expense of disposing this mass in an ecologically safe manner as well as lowering production costs.

The uncertain climate in 1982 regarding the world price of oil kept Venezuela's oil diplomats busy. Calderón Berti insisted upon OPEC adhering to its production-level agreements, while PDVSA moved to improve corporate ties with Mexico's PEMEX and Brazil's PETROBRAS.

PETROBRAS has had its own political problems at home in the politics of energy. The bureaucratic politics of Brazil are no less intense than Venezuela's partisan conflicts. PETROBRAS, NUCLEBRAS (Brazil's nuclear-energy enterprise), and ELECTROBRAS (Brazil's electric-production enterprise) are intensely debating the role of fossil fuels, biomass energy, hydroelectric energy, and nuclear energy. PETROBRAS, through its overseas investment affiliate, BRASPETO, offered Guatemala and Trinidad and Tobago technical assistance in developing their own oil reserves.

There is a feared PETROBRAS-Brazilian intrusion into the PDVSA-Venezuelan Caribbean area. PEMEX in 1981 also outproduced PDVSA. PEMEX lacks adequate storage facilities for its heavy oil, and because Mexico is not a member of OPEC, Mexican oil (three-fourths of which went to the United States) was sold as quickly as possible.

The PDVSA corporate response was to work toward a PDVSA-PEMEX-PETROBRAS compact. In March 1982, PDVSA, PEMEX, and PETROBRAS created a multinational oil company, PETROLATIN, headquartered in Panama,[34] apparently designed to lessen the possibility of conflict between PDVSA, PEMEX, and PETROBRAS. There is to be PETROLATIN activity in Guatemala and Ecuador and the PDVSA has also joined with other Latin American state petroleum companies to create a mutual-assistance program (ARPEL, Latin American State Petroleum Mutual Assistance Program). Colombia's state-owned ECOPETROL and Ecuador's CEPE are also members of ARPEL.

The politicalization of the PDVSA advanced during the Falklands-Malvinas war of 1982. The government in mid-1982 forced the PDVSA to withdraw $2 billion reserves invested in British banks, perhaps the worst sort of politicalization possible. Reliability and nonboycotting for political reasons have been Venezuelan selling points for its oil. Venezuela has insisted that its oil be an instrument of international negotiations and of assuring respect for its democratic development, not of international confrontation.

The problems of PDVSA regarding its new domestic accountability in the 1980s are even more complex. Neither the politicians nor the managers of PDVSA can ignore the fact that all 15 million to 16 million Venezuelans rely on the work of the 42,000 employees of PDVSA and that these people will show their resentment toward PDVSA's privileged few whenever the economy slows down. The role that PDVSA has assumed as a populist scapegoat might be explained by the fact that in April 1982 President Hererra ordered a general austerity program. The price of domestic gasoline was raised for the first time in about 20 years, to 24 cents a gallon for regular and 88 cents for premium.

PDVSA's continuing political dilemma is illustrated by its offshore activities. While the Orinoco Heavy-Oil Belt secures Venezuela's future, the present requires that it find new light oil. The search for light oil has moved offshore. Rather than being free of political controversy (since no one lives in the sea), off-shore oil exploration has complicated the highly emotional maritime border dispute between Venezuela and neighboring Colombia. There was an effort to resolve this dispute in 1980 in order to allow oil exploration in the Gulf of Venezuela, but the suggestion that any oil found in the area would be shared with Colombia caused an emotional outburst in Venezuela. Efforts at seeking a diplomatic resolution to the maritime dispute have, once again, been postponed. The need to seriously conserve the dwindling reserves of light oil motivated the 1982 increase in the domestic price of gasoline. In 1982, although PDVSA announced that it had discovered a small reserve of light oil in waters not subject to foreign claims, the off-shore drilling was disappointing and there was the possibility that it would cease or be scaled down.

PDVSA's critics, such as former Energy Minister Pérez La Salvia, have used Mexico's PEMEX as their model of comparison. MAS leader Francisco Mieres has suggested following Mexico's more politically sensitive PEMEX model rather than the "technocratic concentration with the transnationals" that has characterized the PDVSA.[35] The PDVSA has also been criticized by Mieres as lacking both a "fervor of the masses" and "nationalist sense." It should be noted that the histories of oil nationalization in Mexico and Venezuela have not been comparable. Oil in Mexico in 1938 did not have the same central importance that it has had in contemporary Venezuela. Mexico's resistance to the transnationals' intrigues and boycotts following nationalization was heroic and commendable. However, while the misbehavior of the transnationals in the 1939–41 period could explain PEMEX's difficulties in pre-World War II days, it cannot by itself explain "the 21 years it took PEMEX to show a profit and the 36 years it took before Mexico became a petroleum exporter once again."[36] In 1980, perhaps in response to the fervor of the masses, PEMEX employed over 100,000 workers—more than double the number employed by PDVSA.

The pluralist model of the PDVSA, with its healthy competition among the Four Sisters and its democratic oversight as previously described, rose naturally out of the Venezuelan experience and is well-suited to the Venezuelan style of blending political and managerial impulses. It has produced a unique public enterprise that has been profit-making and politically accountable. Rather than looking toward the experience of other societies such as Mexico, Venezuela should demonstrate a self-confidence in its own organizational creativity and proceed down the road to dominate petroleum with a sense of national pride and accomplishment. Venezuela should rely on its own experience and disregard foreign models and concepts.

The bottom-line considerations regarding the six-year experience of postnationalization are the questions of whether it has continued to make a profit and whether the industry's operations have been accountable to both the government and the nation. The answer in both cases is emphatically yes. Between 1976 and 1982, PDVSA taxes paid to the nation's treasury more than doubled. Revenue from oil sales rose despite the need to reduce production to dry up the oil glut. PDVSA's after-taxes profit in 1981 was $3.1 billion. Part of the 1980 and 1981 profit was from interest earned from its vast reserves and investment funds. More importantly, the nation's proven reserves in crude oil increased by 12 percent after many years of decline. Natural gas reserves increased by 35 percent, and new light oil was discovered.[37]

The political accountability of the PDVSA is a most difficult aspect to discuss. In the 1983 period of austerity, the relationship between the PDVSA and the nation is being restructured. On one hand, there is no desire to have the PDVSA become as *criollo* as the state-run telephone company. That would be a disaster. On the other hand, pressure to deepen the meaning of the 1976 nationalization is growing. A democracy cannot ignore the anguish of growing numbers of unemployed workers, disillusioned university graduates, and bankrupt businessmen.

THE FUTURE OF OIL

The now-classical definition of a dependent, enclave economy was given by Charles Anderson. According to Anderson, the benefits of the enclave's exports

> Only peripherally and almost accidentally touched the nation. . . . All processing of raw material was done abroad. . . . Capital and management was provided almost entirely by foreigners. *Profits were either reinvested in the enclave or carried back to the country in which the enterprise originated.*[38]

Since 1936, Venezuela has moved cautiously but definitively to erode this dependency. The 1976 nationalization secured the profits and the management of the oil industry for Venezuela. The exotic export-oriented petroleum sector had been allowed to remain largely aloof from the remainder of Venezuela's national-exchange economy in the aftermath of nationalization in the 1976–82 period.[39] The operating affiliates of the PDVSA even maintained their former intimacy with "their" transnationals by means of technology and marketing contracts. This may have been necessary. The PDVSA received outstanding marks for its excellent maintenance of the industry's profitability in a time of transition and in keeping the bridges open to New York and London.

The economic recession of the early 1980s, however, made it necessary to build a broad, two-way bridge linking the PDVSA to the rest of Venezuela. To paraphrase the words of COPEI Sen. Valmore Acevedo, Venezuela is more than the owner of an oil company; it is a nation. It is also a democratic nation that takes more than company profits or technical planning facts into consideration in deciding public policy.[40]

A top PDVSA official reflected, with some poignancy, that the state oil company was perceived by many to be some benign, foreign-oriented Gulliver, bearing the gifts of rational management and technical proficiency to a nation of culturally backward Lilliputians. This was not a favorable nor overly optimistic perception of the relationship between the modernized management of the company and the culturally bound political sector. The example of the 1975 bankruptcy of Indonesia's state-owned oil company, PERTAMINA, illustrated the fallacy of overreliance on one "island of managerial rationality" to solve a nation's problems.[41] In seeking to better integrate company and nation, what would be the trade-off between profitability and accountability?

The Falklands war of May 1982 was used as a cover to begin the political restructuring of the PDVSA. The need to demonstrate Latin American solidarity and to take a strong anticolonialism stance were the ostensible reasons for repatriating $2 billion of the PDVSA's international reserves held in British banks. A more convincing argument could be made, however, that the Falklands were only a cover for the government to move to gain control of the entire $8 billion in the PDVSA international reserves. Venezuela's total international debt in early 1983 was about $30 billion. The myriad of mismanaged public enterprises alone had an international debt of $18.5 billion.[42] The reconversion of this short-term debt into a long-term one was apparently facilitated by the repatriation of the PDVSA's funds and its being placed under Central Bank control.

The PDVSA will now have to seek external funding for new, major investments, such as the full-scale development of the Orinoco Belt, but there are many Venezuelans who favor having the international community share in the risks of this innovative heavy-oil project. AD and its former Energy Minister Valentine Hernández are strong proponents of an innovative use of controlled foreign capital in the Orinoco and of Venezuela buying into international service companies.[43]

There are many astute Venezuelans who see great opportunities in the current fiscal crises to truly integrate oil and the nation and to impress upon the population the need for a serious economic diversification. The period of petrobonanza (1974–80) was filled with many pseudo "great leaps forward." During this period, vast sums of money were spent on many ill-considered "megaprojects" with little thought given to the very real problems of implementation and maintenance. Complex, pluralist democracies often need the challenge of a crisis to achieve a creative consensus for change. The principal features of the emerging structure of the PDVSA relationship to the rest of Venezuela in early 1983:[44]

National procurement of material goods and professional services as long as quality and reliable supply are assured. A reasonable cost penalty should be tolerated in order to stimulate the Venezuelanization of the PDVSA's procurement. This preference for Venezuelan products was already in process, and between 1978 and 1982, the percentage of national content in the company's procurement doubled, reaching 40 percent in 1982.

The technological contracts between the PDVSA and foreign companies should be more closely monitored. Foreign contractors are now expected to establish Venezuelan training centers and to accept the eventual Venezuelanization of the technology.

The PDVSA has been definitively discouraged from its overreliance on the transnationals for marketing its oil. Greater reliance on long-term government-to-government arrangements and linkage of oil sales to technology transfers is underway. About 25 percent of the global trade in oil in 1981 was by means of government-to-government arrangements. Venezuela has actively found new markets and is less dependent on the U.S. market.

The company has been drawn into the membership of two interministerial commissions assigned to revitalize the Lake Maracaibo area development. The principle that oil is to develop a balanced and integrated nation, not just an enclave, is in operation. About $1.1 billion of PDVSA funds have been allocated to the urban redevelopment of communities on the east coast of Lake Maracaibo.

The antagonistic blending of the political and managerial impulses in seeking a new relationship between oil and the nation was illustrated by the PDVSA's being encouraged in late 1982 to purchase $1.75 billion worth of bonds in what was Venezuela's largest private bank, the Bank of the Venezuelan Worker (BTV). The BTV was created in the mid-1960s with the Labor Confederation (CTV) as its principal shareholder and with considerable government financial support. Many in the CTV saw in this union-owned bank the basis to make labor unions autonomous from the government and the source of innovative industrial structures of worker-owned and worker-comanaged enterprises. During the years of the petrobonanza, the BTV expanded rapidly and apparently overextended itself. By 1982, it was Venezuela's largest private bank, with 10 percent of total deposits, but it was also failing to meet its own debts.[45] The COPEI government intervened and encouraged the PDVSA to make a sizable investment to shore up this bank. There was concern that the competent management and scarce funds of the PDVSA would be overstretched to accommodate short-term government social and financial difficulties, distracting from its effective operations as an oil company.

Venezuelan oil diplomacy has been able to lessen the disadvantages of its increasing dependence on heavy oil. The newly built refineries in the Caribbean basin can blend Venezuelan heavy oil with lighter oil to produce both gasoline and other petroleum by-products. In late 1982, Venezuela and Nigeria agreed to swap equal amounts of Venezuelan heavy oil and Nigerian light oil to even out somewhat the sales mix of these two OPEC nations. Venezuela has also succeeded in diversifying its markets and in reforming its sales, from an industry nearly totally dependent on transnationals to one involving long-term government-to-government swaps and agreements. Venezuela's tradition as a friendly, reliable supplier has also enabled the nation to keep its share of its traditional markets and to suffer less of a price decline than other OPEC members.[46]

NOTES

1. Franklin Tugwell, "Venezuela's Oil Nationalization, The Politics of Aftermath," in *Contemporary Venezuela and Its Role in International Affairs*, ed. Robert D. Bond (New York: New York University Press, 1977), p. 92.
2. Annibal V. Villela and Werner Baer, "Evaluating Public Enterprises: Some Dilemmas in Formulating Guidelines," (First international seminar on Public Enterprises and Economic Development, Brasilia, July 27–31, 1981), p. 1.
3. "Gov't Wants More Power for Business," São Paulo, Brazil, *Daily Post*, July 31, 1981.
4. Villela and Baer, "Evaluating Public Enterprises," pp. 4–5.
5. Interview, Senate Office, Caracas, August 1980.
6. *Latin American Economy Report* 5 (July 1, 1977): 1000.
7. "Nuevo Presidente del Banco Central de Venezuela," *SIC* 44 (April 1981): 81.
8. Tugwell, "Venezuela's Oil Nationalization," p. 110.
9. Gen. Rafael Alfonzo Ravard, "The Venezuelan Oil Industry and Its Investment Plan (1980–1990), in *Venezuelan Development: Actual Situation and Perspectives* (Presentations made at the annual joint assembly of the International Monetary Fund and the International Bank for Reconstruction and Development, October 1980), p. 89.
10. Interview with a PDVSA executive, Caracas, December 1981.
11. "Dijo el Presidente de PDVSA: Logramaos la Normalizacion de la Industria Petrolera," *El Nacional*, December 20, 1978.
12. Kim Fuad, "Oil Operators: The Forgotten Foursome," *Business Venezuela* (November/December 1981), pp. 27–32.
13. Tugwell, "Venezuela's Oil Nationalization," p. 99.

14. Peter Greiff, "Attaching Strings to Oil Sales," *Business Venezuela* (May/June 1981): 69.

15. "Venezuela: PDVSA in West Germany Deal," *Latin America Regional Reports: Andean Group* (January 28, 1983).

16. Humberto Calderón Berti, "Energy, An Opportunity for Hemispheric Cooperation," Embassy of Venezuela *Venezuela Up-To-Date* 21 (Spring/Summer 1980): 27.

17. Julio Sosa Rodríquez, "U.S. and Venezuela: The Oil Diplomacy" *Latin American Times* 3 (March 1981): 9. See the author's previous discussion on this issue; David E. Blank, "Venezuela: Politics of Oil," in *U.S. Influence in Latin America in the 1980s*, ed. Robert Wesson (New York: Praeger, 1982).

18. Celestino Armas, "El Articulo Quinto," *El Nacional*, August 7, 1981.

19. Interview, Caracas, August 1980, José Giacopini Zarraga, personal advisor to the president of PDVSA, and "50 por Ciento del Petróleo So Vendera a Nuevos Clientes," *El Nacional*, August 4, 1980.

20. Tugwell, "Venezuela's Oil Nationalization," pp. 94–95.

21. Octavo Lepage, "Las Grandes Decisiones en Materia de Petrolera No Han Sido Tomadas por los Tecnicos," *El Nacional*, April 22, 1981, and "Se Acaba el Tabo Petrolero," *SIC* 44 (May 1981).

22. "COPEI Respalda el Manejo de la Industria Petrolera," *El Nacional*, April 22, 1981.

23. "Heavy Oil Project Sparks Controversy in Venezuela," and "Petroleum Flows in Orinoco Valley Like Blackstrap Molasses in January," Washington *Post*, January 27, 1980.

24. "Calderon Berti Says Time Has Come to Extend Oil Industry's Contribution to Development," Caracas *Daily Journal*, July 24, 1980.

25. U.S. Senate Committee on Energy, *The Geopolitics of Oil* (Washington, D.C.: Government Printing Office, 1980).

26. "Bechtel: A Reclusive Grant," New York *Times*, July 8, 1982.

27. "Conflict at Venezuelan Oil Company," *Wall Street Journal*, February 16, 1982.

28. "Conflict at Venezuelan Oil Company," *Wall Street Journal*, and "Profusa Actividad Petrolera," *SIC* (September-October 1981).

29. "PDVSA Profundiza las Divergencias Oficiales," *El Diario De Caracas*, December 27, 1981, and "Enter Díaz Bruzual, Bearing Hatchet," *Latin America Weekly Report*, January 8, 1982.

30. "La Polemica, PDVSA-BVC," *El Diario De Caracas*, December 27, 1981.

31. Hugo Pérez La Salvis, "Venezuela Carece una Verdadera Industria Petrolera," *El National*, December 27, 1981.

32. "Enter Díaz Bruzual," *Latin America Weekly Report*.

33. Kim Fuad, "Heavy Crude, Hope for Tomorrow," *Business Venezuela* (January/February 1982): 33.

34. *Latin America Weekly Report*, March 1, 1982.

35. Francisco Miéres, "El papel del petróleo venezolano en el perspectiva de la crisis energética," in *Petróleo y Desarrolla en México y Venezuela*, Marcos Kaplan, ed., (México: Universidad Nacional Autónoma de México, Editorial Nueva Imagen, 1981), p. 233.

36. Shoshana B. Tencer, "OPEC and Petroleum Policies: The Growth of Government Controls over Exports of Raw Materials," (Paper delivered at the Latin American Studies Association meeting, San Francisco, November 14–17, 1974).

37. Kim Fuad, "PDVSA: A Bottom Line to Boast About," *Business Venezuela* (March/April 1982): 35.

38. The author wishes to thank Professor Ken Coleman, University of Kentucky, for focusing on the need to achieve national integration and to cite Charles W. Anderson, *Politics and Economic Change in Latin America: The Governing of Restless Nations* (Princeton, N.J.: Van Nostrand, 1967), quoted in Ken Coleman, "Self Delusion in U.S. Foreign Policy: Conceptual Obstacles to Understanding Latin America," (Monograph 3, Northwestern Pennsylvania Institute for Latin American Studies, Mercyhurst College, Erie, 1977).

39. Coleman, "Self-Delusion."

40. Interview with COPEI Senator Valmore Acevedo, Caracas, December 1981.

41. Willard A. Hannah, "The PERTAMINA Perplexity," *American Universities Field Staff, Asia* 24.

42. "Venezuela Seeks Debt Conversion," New York *Times*, September 30, 1982.

43. Kim Fuad, "Foreign Role in Oil," *Business Venezuela* (July/August 1982): 13.

44. "Quarterly Economic Reviews: Venezuela," *The Economist* (1st, 2nd, 3rd quarters, 1982).

45. "Venezuela Seeks Debt Conversion," New York *Times*, and *Latin American Weekly Report*, December 1982.

46. "Venezuela: Facing the Petroleum Glut," *Latin American Times*, November 1982, p. 7.

9

MANAGEMENT AND POLITICS: INDUSTRIALIZATION, PARTICIPATION, AND DECENTRALIZATION

Venezuela's fundamental development goal has been to create an autonomous economy and a socially just and integrated society. The oil sector, left to itself, traditionally distorted the economy by creating an isolated export-oriented sector and a weak, national-exchange industrial and agricultural sector. Oil, despite its financial contributions, also fragmented Venezuelan society. The politics of "sowing petroleum" have been the politics of lessening Venezuela's dependence on that export commodity.

This chapter is an effort to analyze the performance of Venezuela's pluralistic politics as Venezuela confronts the challenges of industrialization and the concomittant search for social justice. Three concrete case studies illustrating the politics of industrialization and enhancement of social justice have been selected to demonstrate the variety of political styles operative in Venezuela due to its mix of managerial and political impulses. The utilization of case studies also enables one to penetrate the layer of theory and rhetoric that often impedes an examination of the real-world progress of pluralist rule making and implementation in a developing nation. The politics of industrialization have focused on the allocation of resources within the framework of a broad national consensus. The politics of industrialization have primarily been confined to the nonpartisan planning sphere. As industrial-planning decision making has, however, increasingly affected jobs and labor-security issues, the ability of planners to insulate their activity from the electoral-based partisan politics has accordingly diminished.

INDUSTRIALIZATION: THE AUTOMOTIVE INDUSTRY

Following the bloodless and instantaneous military overthrow of the Rómulo Gallegos government in 1948, Democratic Action (AD) party leaders were impressed with the difficulty of transforming the country by means of revolution and radical reforms. By the time Rómulo Betancourt was inaugurated as president in 1959, industrialization was accepted as the principal means of achieving a just and modern Venezuela. Claudio Véliz explained the process by which industrialization replaced radical reform as the core of the middle-class ideology in Latin America:

> The urban middle class . . . had a vague hope that industrialization would perform painlessly and automatically the social transformation they had not been able to bring about themselves. Industry became a gigantic *Deus ex machina* expected to solve social and political problems, and change attitudes and bring prosperity.[1]

Professor Véliz was not writing specifically about Venezuela, but one would only have to add words such as "to lessen the nation's dependence on oil" and "to peacefully amalgamate modern and traditional values" to fully appreciate just how applicable to Venezuela was his 1965 essay. One need only read the six national plans that have provided a basic guideline to Venezuela's aspirations to understand the critical importance of industrialization to Venezuela's post-1958 politics. It was quite audacious to think that Venezuela, a nation of only 8 million people in 1960 (the majority of which were illiterate), could have a viable automotive industry.[2] With all its defects and vices, democratic populist muddling-through has achieved just that. In 1980, Venezuela's automotive industry proved itself to General Motors, Ford, Nissan, and Renault.

The politics of industrialization and the "motors war" over the automotive industry have been called an "intense intrabourgeois conflict" by one group of scholars, implying that the basic conflict was between those business interests developing an export capability and those focusing their sales almost exclusively on the national market.[3] The politics of the automotive industry, because of the importance given to that industry, have also had a regional and global aspect. On one hand, Venezuelan populist politicians have intervened in the industry's planning in order to deliver an automotive-parts or assembly plant to their community. On the other hand, Venezuela has had to relate to the global strategies of the transnational automotive firms and to the competitive development aspirations of other Andean Pact members.

The pivotal position assumed by the automotive industry in the overall industrialization program was due in part to the existence of oil and the discovery in 1941 and 1947 of two large iron-ore deposits. The tremendous hydroelectric potential of the Orinoco-Caroni River system had also become known. With these basic raw materials, Venezuelan planners aspired to transform their nation into an automotive power. Fear as well as hope motivated Venezuela in the 1945–58 period. The atomic age had begun, and there were indications that nuclear-power plants might shortly replace oil-fired furnaces to heat the homes of Europe and the United States. The United States first indicated in 1954 that it might limit its oil imports in a time of world overproduction.

There was both a sense of urgency and a broad national consensus regarding industrialization. Indicative of this consensus, the plans of the *trienio* AD government (1945–48) to industrialize the frontier Orinoco Valley (Guayana region) were first implemented by the Pérez Jiménez dictatorship. The beginning of a Venezuelan automotive industry, albeit one focused exclusively on import substitution, also occurred during the 1950s. By the time constitutional government was restored in 1958, government-owned steel and petrochemical complexes were under construction, and two foreign-owned assembly operations (General Motors and Chrysler) had been established. Auto parts such as tires, exhaust pipes, and batteries were also being produced by subsidiaries of U.S. companies. In 1962, the Central Office of Coordination and Planning (CORDIPLAN) published a policy outline for the development of a Venezuelan automotive industry.[4]

The CORDIPLAN report set the 1970 programming targets: the creation of an integrated import-substitution automotive industry of more than 70,000 vehicles assembled per year, 60 percent of whose net weight would be Venezuelan-made parts. An estimated 40,000 workers were projected for the industry in 1970, compared to the 1,800 employed in 1961. This news was well-received in Venezuela, which in the early 1960s had over more than 13 percent unemployment as the oil boom of the 1950s ended.

Venezuelan government planners, with the help of experts from the U.N. Economic Commission for Latin America, envisioned that the decision-making process for the automotive industry would be contained within a mixed government-private sector programming group. This group in the late 1960s was hailed as an example of concerted or indicative planning.

The primary purpose of the group was to maintain a narrow range of conflict, allowing transnational firms, private Venezuelan parts manufacturers, and government agencies to plan the industry's development free from extreme partisan conflict and ideological disputes. In order to exclude problems relating to production costs and profits, progress in the industry was to be measured by the substitution of foreign-made parts and components by Venezuelan-made ones, and this was to be measured in net weight rather than in value-added. Labor unions were therefore not to be a factor in planning the industry.

Government planners envisioned only three actors in the programming group: the concerned government agencies—Ministry of Development State Development Corp. and CORDIPLAN—the transnational assembly firms, and private-sector national-parts manufacturers. In order to maintain the planning dialogue concerned with broad sectoral issues, the two private-sector groups (the transnationals and national parts makers) were required to form industrial associations to represent them. The transnationals formed the Chamber of Automotive Assemblers, and the nationals formed the Venezuelan Association of Automotive Parts Manufactures.[5]

Between 1962 and 1970, the programming group succeeded in creating a reasonably successful national auto industry. The basic targets of the 1962 CORDIPLAN report were essentially achieved, but the quality and costs of the automotive vehicles assembled left much to be desired. While 33 percent of the assembled vehicles were nationally made parts, all the key and costly components, such as engines, brakes, and drive trains, were still imported. Only relatively simple "hang-on" parts were truly Venezuelan. The progressive Venezuelanization of the industry was being blocked by the proliferation of models and makes assembled, which prevented mass production of parts and scale of economy. In a sense, Venezuela was being colonized by the transnational assemblers without gaining much employment and technical know-how. For example, in 1965, 13 different firms assembled no less than 64 models of vehicles for a national market of fewer than 44,000 units.

By the late 1960s, Venezuelan planners changed the rules of the game. The industry was to be redesigned to become a potential export industry. On one hand, Venezuela began the negotiation process that led to its joining the Andean Pact in 1973. The Andean Pact required using value-added rather than net weight to judge national content. The second transformation of the industry was that the basic engine blocks and drive trains would have to be made either in Venezuela or in another Andean Pact country (Colombia, Peru, Ecuador, or Bolivia). Decision making became too complicated for a contained sectoral-planning dialogue. The basic policy instrument for planning the industry in the late 1970s was not entirely a Venezuelan document but was Andean Pact Decision 120, finally approved in 1977.[6]

Under Decision 120, Venezuela's automotive industry was radically transformed. In addition to receiving Andean Pact allocation of engine manufacturings, the industry is now intimately linked to Guayana Corp. (CVG) steel and aluminum production. General Motors (GM) is one transnational that believes Venezuela "can achieve significant comparative advantages . . . with the production of aluminum parts."[7] GM bought out the failing Chrysler operation in 1979 as a means of showing its faith in Venezuela. Venezuela's contemporary, basic automotive policy was the result of its 1973 entrance into the Andean Pact. Unlike the 1962 CORDIPLAN policy, the 1975 policy document required assistance from the Foreign Commerce Institute, created to spur exports via the Andean Pact. The central target of the 1975 decree was the local production of 90 percent of the vehicle's value, including the drive train by 1985.[8] The 1975 policy also called for Andean Pact coproduction, in part to avoid any one member from being colonized by one transnational and to encourage the production of a hybrid car that would be a uniquely Andean-Pact product.

The hybrid-car concept meant that a Pegaso diesel engine could be used in a Mack assembly along with a GM exhaust system. When the prestigious Mercedez-Benz assembly operations in Venezuela resisted hybridization it was eventually invited out of Venezuela and the Andean Pact market. This may have enhanced its desirability, but future purchasers of Mercedez would have to pay either substantial tariffs or, most likely, substantial bribes to import them.

While in 1978 the Venezuelan government conceded that achieving 90 percent local-parts value integration was not realistic, about 75 percent of the value of the cars produced in 1981 was local content. Venezuela also assembled 182,000 automotive vehicles that year. Venezuela in the late 1970s was the target of many high-powered visitors as the scramble for a share of the Andean Pact-Venezuelan automotive industry heated up. Henry Ford, aided by Nelson Rockefeller, made a well-publicized visit to Caracas in order to dispute a potential GM and Renault domination.[9] King Juan Carlos of Spain visited Venezuela in 1977 to personally help the cause of the ailing Spanish automotive industry.[10]

By 1981, Venezuela had created a viable automotive industry that employed 15,000 workers in assembly operations and an additional 30,000 in the production of parts and components. The investment in the parts industry is largely national, although a few large family groups dominate. The quality of Venezuelan automotive products has also vastly improved. Venezuela accounts for more than 70 percent of Andean Pact market production.

The Social Christian Party (COPEI) administration in the late 1970s and early 1980s was seeking to make the industry more competitive, focusing on the nation's comparative advantages in aluminum, steel, and petrochemicals rather than on high protective tariffs and government overregulation. However, while COPEI believes that in the longrun the industry's rationalization will make it competitive in the Andean Pact and global markets, in late 1982, import restrictions were strenghtened in fear of Andean Pact competition.

One of populism's intrusions into the industry was President Pérez's granting of "exclusive" rights to produce heavy diesel engines in Venezuela to both Spain's Pegaso and the U.S. Mack Co. The Andean Pact would allow only one such plant in Venezuela.[11] While some Venezuelans continue to fear the influence of the transnationals in the industry, suggesting that industrialization has only reproduced dependency, this writer believes that the creation of a viable auto industry, producing close to 200,000 vehicles a year, in 20 years has been a remarkable achievement. In fact, there is evidence to suggest that the transnational auto makers have not made much money from their Venezuelan investments. This industry's progress seems to confirm that the proper blending of the managerial and political impulses that is the mark of a pluralist democracy such as Venezuela need not inevitably be stalemated or immobilized.

Labor in the automotive industry is also well-organized and well-remunerated. Next to the petroleum workers, auto workers are Venezuela's best paid. Labor's intervention in the planning of the industry has been largely defensive, to make certain that the workers are not the only ones paying for the cost of progress. In 1964, Ford Motors tried to hold its workers hostage in order to violate the programming group's import-substitution-of-parts policies. Ford's threats to lay off 900 workers if it did not get its way led to a major political crisis in 1964 and to direct presidential intervention in the industry's planning. Ford accepted the government's compromise formulas.[12]

The labor unions have joined with the domestic business sector to express their concern that with gasoline prices rising, there is the potential, under the Andean Pact, that cheap and small Colombian cars will flood the market, while the bigger Venezuelan cars will find no buyers in the poorer economies of Colombia, Peru, and Ecuador.[13] This fear led in late 1982 to the imposition of import restrictions on more than 150 items. This unilateral Venezuelan action has threatened the future of the Andean Pact.

Labor is also a member of the trilateral (government-business-labor) commissions that oversee many industrial policies, such as the issue of industrial labor discipline, health and safety, and the introduction of technology. Populism has given the industrial sector a labor code that makes the dismissal of workers for cause most difficult. The law on unjustified dismissals requires the payment of two months salary for each year of employment. Labor discipline is allegedly almost nonexistent, a fact aggravated by the continuing shortage of skilled industrial labor. "Venezuelan labor legislation including matters such as compulsory overmanning, minimum wages, obligatory salary raises and most of all the law against unjustified dismissals, is regarded [by managers] as even more disastrous than price regulation."[14] Writing in 1979, a European scholar stated this management position. "The car assembly industry has been working at a loss for years. . . . The fear of loss of investment and the hope for a drastic increase in sales to the Andean Common Market serve to explain why the large automobile multi-nationals are not closing down their operations in Venezuela."[15]

Populism's intrusion in industrial planning also means that after a company, especially a foreign-owned one, has reached an agreement with a labor union affiliated with one political party, a second labor union, affiliated with another party, can come along to offer a better deal. Workers have been known to join several populist political parties in order to enhance their influence, the policy of "parallel unions."

In addition to securing a modern and autonomous Venezuelan economy, industrialization and the state intervention in agriculture and industry were to create a participatory and socially just Venezuela. Before discussing industrial labor and peasants, a few words on the ability of populism to do away with the myth of marginality are in order. Venezuela's social prestige and economic wealth remain inequitably distributed, but populism, with all its defects, has empowered Venezuela's workers, peasants, and urban poor. As has been indicated, Venezuela's workers and peasants are an integral, if somewhat exploited, part of the national reality and are not locked into any self-defeating and pathological culture of poverty.[16]

The populist political impulse since 1958 has promised more and perhaps delivered less to the nation's peasants and industrial workers than to any other group. This is not intended to be a criticism of Venezuela's political-party system, nor is the author charging democracy with failing at agrarian reform or social justice. Rather, there is a recognition in all democratic societies that creating effective means of popular participation and creating the instrumentation for equality are most difficult tasks.[17]

POPULAR PARTICIPATION AND SOCIAL JUSTICE: INDUSTRIAL LABOR, AGRARIAN REFORM

Industrial Labor

The Venezuelan Confederation of Labor (CTV) has its origins in the 1936–48 decade of transition. The first national Congress of Workers took place in December 1936. During its early years, Venezuela's labor movement was about evenly split between unions controlled by Communist (PVC) and AD leaders.[18] AD's participation in the October 1945 overthrow of President Medina and its control of the government of the *trienio* (1945–48) led to AD's hegemony within the CTV.

One critical account of CTV leadership, written in 1981, called it "nothing, but an instrument of AD; the other parties exist in it merely to legitimize AD's domination"[19] and to get some of the patronage and payoff. COPEI tried to create its own, parallel, labor confederation, the Confederation of Autonomous Unions (CODESSA), but it has been largely unsuccessful. CODESSA divided in 1971, giving birth to the General Confederation of Labor (CGT). In 1961, as part of the growing schism between the old-guard AD leadership and the youthful radicals of the Revolutionary Left Movement, the Communist Party and the Democratic Republican Union were expelled from the CTV. These antigovernment labor leaders formed the Unified Labor Central of Venezuela (CUTV). The CUTV has splintered several times, and like CODESSA-CGT unions, it has an insignificant affiliation. MAS, the Eurocommunist party (it prefers to be called Socialist), has focused its labor-union activity in the CTV.

The domination of the CTV in the labor movement is illustrated by the following data: Of the 1,170,000 estimated organized workers in Venezuela, 1,000,000 belong to CTV-affiliated unions. In the eighth CTV congress, held in October 1980, AD won nine of 17 seats on the executive committee. COPEI, largely with AD's acquiescence, was ceded the second position, and perhaps in deference to the 1978 presidential elections, CTV Secretary Gen. Rafael León is a COPEI member. The People's Electoral Movement (MEP) clearly retained the third force in the CTV, beating back a rigorous MAS challenge. Indicative of the fact that CTV's politics reflect partisan politics, part of the AD-COPEI 1978 clash was reflected in the division of the once-great Democratic Republican Union (URD). In 1980, URD was generally supportive of COPEI, and the MORENA party of Leonardo Montiel Ortega, which broke from URD, was largely supportive of AD; URD and MORENA have insignificant support in the labor movement. Yet an AD-COPEI compromise in the 1980 congress awarded each one seat on the CTV Executive Committee.

Industrialization is relatively new in Venezuela. Most workers labor in work places that employ fewer than five people, and they have constituted the nonorganized artisan sector. Of the 27,000 private-sector firms that employ more than five people, only 3,150 can be considered big business, employing more than 50 people. It is this big-business sector that is largely unionized. About half the industrial jobs created over the past decade have been created in the public sector. In many ways, the role CTV has played in labor-management relations in the public sector has been the cause for its possible transformation.

The CTV leadership structure is highly centralized. Local-union grass-roots autonomy is an issue that is now being heatedly contested. The CTV-AD/COPEI leadership has two means of intervening in local union affairs. On one hand, all CTV locals belong to an industrial federation, while on the other, they are also part of a state labor federation. The key battleground for the future of Venezuelan labor is the struggle to control the nation's largest local union, the Steel Workers and Allied Activities Union (SUTISS).

SUTISS is the recognized local of 16,000 members in the state-owned Orinoco Steel Co. plant in the planned industrial city of Ciudad Guayana. Life for youthful industrial workers in this new frontier town has been difficult. Work in a tropical steel plant is also quite dangerous and extremely uncomfortable. A long history of both community-related and work-place-related grievances has made SUTISS a traditionally active union. In 1979, *Causa R* (Radical Cause), led by Alfredo Maneiro, organized a faction of labor leaders that swept the SUTISS elections. Causa R won four out of the ten SUTISS leadership positions, while four other allied, radical factions—the Trotskyist-Socialist League, the Popular Struggle Committee, MIR-AMERICO, and the Radical Action Group (Catholic left)—won one seat each. The two populist, status parties—AD and COPEI—each won only one seat.[20]

Orinoco Steel Co. (SIDOR) has been a troubled state enterprise, with huge losses and repeated problems with corruption. SIDOR losses in 1981 alone approached a half-billion dollars. It has also been a critical link in Venezuela's development strategy. SIDOR's production of flat, rolled steel and forged steel blocks is a vital input for the Andean Pact automobile industry. There was a fear that honest labor leadership in SUTISS would both uncover a history of "sweetheart" agreements with the state-owned company and make costly new demands that could threaten the viability of the generation-old steel-town development.

There is no doubt that excessive employment of redundant managerial and production workers is a key factor in SIDOR's problems. SIDOR's production capability was shattered by the decision of the AD administration (1973–79) to attempt to rapidly quadruple its output. Under the *IV National Plan,* massive investments were poured into SIDOR. These were uncoordinated and poorly executed. The company's good managers were swamped with new responsibilities, and the poor managers were overwhelmed by the new opportunities for graft. The plant and the planned industrial city broke down in the late 1970s, overwhelmed by the boom-town reality.

The AD/COPEI-dominated CTV moved to "intervene" SUTISS in 1981, thereby ending its local autonomy. The CTV leadership utilized both the industrial Metal Workers Federation and the state federation to bring about its intervention.

The future of state enterprises and the autonomy of the labor movement may be at stake in the SUTISS issue. There is evidence to suggest that populism's future viability is also being determined. The struggle for SUTISS has caused mass demonstrations and a gang-land resort to praetorianism. Gangs of armed men have apparently been utilized by the AD faction in the SUTISS struggle.[21] The editor of the respected magazine, *Resúmen,* who had previously revealed former President Pérez's alleged involvement with criminal elements and the left-leaning Jesuit, Centro Gumilla, have both taken up the cause of SUTISS autonomy. Jorge Olavarría, Resúmen's editor, was considering running as an independent or left-center unity candidate in the 1983 elections.

Agrarian Reform

Venezuela's agrarian reform celebrated its twentieth anniversary in 1980. This event caused a flurry of press and scholarly work on the balance of this 20-year challenge. In answer to the bottom-line question of who has benefited from the March 1960 Agrarian Reform Law, this writer would have to suggest that no simple answer will suffice. Most observers would accept the 1969 statement of Raul Alegrett, director of the National Agrarian Institute (IAN), that the "antagonistic concepts" and "contradictory provisions" that were made part of the 1960 law should explain some of the ambiguity felt toward agrarian reform.[22] On one hand, the law promised a "transformation" of the agrarian structure, while on the other hand, large landholdings "which performed a social function" would be exempt from possible redistribution.[23]

There is evidence that the populist political impulse never intended agrarian reform to transform Venezuela's countryside. Even before the broad-based coalition drafted the 1960 legislation, President Betancourt moved decisively to prevent autonomous peasant activity that posed a threat to political tranquility. On May 7, 1959,

Betancourt directed that one large farmholding belonging to a Pérez Jiménista official, Hacienda San Juan de Dios, be distributed to small farmers, with the "triple purpose of aiding the landless, seeing that the land was worked, and winning political support for Acción Democrática."[24] However, when a group of peasants occupied similar Pérez Jiménista land without government approval on May 26, apparently led by people not part of AD, Betancourt moved vigorously to remove the peasants and to punish the "instigators."

This pattern of rewarding loyal peasant groups, AD or COPEI electoral clientele, and punishing those who strayed from the approved populist fold was to become a recurring theme of agrarian policy. This was especially noticed in the 1962 struggle between Venezuelan Peasant Federation (FCV) President Ramon Quijada and President Rómulo Betancourt. Quijada broke with AD's old guard to join in the party's second division in January 1962. Once the tremendous financial, bureaucratic, and coercive resources of the government turned against the Quijada group, ostensibly comprised of the grass-roots leaders of the peasants, its power evaporated. Quijada ended up supporting Uslar Pietri in the 1963 elections, but his boast of delivering 400,000 peasant votes was an empty sham.

In 1977, AD President Pérez apparently sided with commercial agriculturalists in areas disputed with peasant organizations. One scholar cited the fact that the AD minister of Agriculture was allegedly a "millionaire cattleman with several huge ranches and farms in Zulia."[25] According to this scholar, the Pérez administration sided with commercial farmers when they invaded communal land and forcibly evicted peasants, with the help of local police and national-guard personnel. When the AD governor of Apure, a rich agricultural state, protested "the abitrary expulsion of peasants [campesinos]" he was replaced by an individual associated with large, commercial, farm interests.[26]

The research of scholars on agrarian reform in Venezuela is replete with the sense that the program has been implemented essentially to give symbolic rather than real reform and that the program "has benefited mainly local party leaders," enhancing their particularistic patron-client control over their peasant followers.[27]

Scholars also seem to be in agreement regarding the land distributed over the past 20 years to an estimated 150,000 peasants. The quality of the land was marginal, and it was often distributed in lots too small to become productive. Even when reasonably good land was granted, adequate technical assistance, credit, and support facilities were not provided. Basically inequitable control of agricultural land and the fact that less than 2 percent of the farms (the larger units) control more than 70 percent of the land remain unchanged.[28]

Estimates made in 1980 indicated that Venezuela's agrarian reform cost nearly $5 billion between 1960 and 1980. "Yet the value of crops produced by the Agrarian Reform farmers has declined" (between 1967 and 1975).[29] The isolated, individualistic, small land grant has failed and is being gradually replaced by peasant cooperatives and peasant enterprises. Peasant enterprises process as well as grow food, and they resemble normal communities with technicians, doctors, and teachers as well as peasants. The basis for both peasant cooperatives and peasant enterprises is collective rather than individual land distribution.

Peasants, according to Raul Alegrett, are no longer living in neglected isolation but in planned communities of from 200 to 500 families with the amenities of urban society: schools, electricity, and health care.[30] Most agree that the future of Venezuela's peasantry is in these peasant cooperatives and peasant enterprises. However, according to a 1976 joint study of IAN and FCV, only about 5.5 percent of peasants were members of peasant enterprises. A full 31.1 percent of reform beneficiaries belonged to no organized group. Table 9.1 essentially summarizes the balance of agrarian reform 1960–75.

Table 9.1. Index of Participation (1976), Agrarian-Reform Beneficiaries (in percentages of total)

Level of Participation	*Type of Program*	*Percentage Total*
Integral community life	Peasant cooperatives or peasant enterprises (collective land)	5.5
	Agrarian centers	2.1
Political brokerage available to individual peasants	Credit union of service unit	18.0
	Peasant League membership	51.5[a]
No organizational membership	Isolated peasants[b]	31.1

[a]Not exclusive of other participatory activities.
[b]May not have yet received clear title to land.
Source: "Les Grandes Temas, Reforma Agraria: Balance 20 Años de una Desafio," *El Diario de Caracas,* August 10, 1980, p. 10.

Part of the reason for the low level of peasant social, political, and economic integration despite the expenditure of $5 billion over the past 20 years is what George Schuyler has called the "intrusion of party loyalty over community progress . . . which has further eroded grassroots democracy."[31] Venezuelan voters have no control over individual candidates but must vote the impersonal party ballot. Until 1979, even municipal councils were elected on this impersonal ballot, along with all other legislative bodies. Ambitious politicians quickly learned that their careers were advanced by gaining influence in Caracas, not by representing a rural community.

The AD administration 1974–79 was publicly willing to admit that agrarian reform was a failure. The old-guard party leaders, led by the late Rómulo Betancourt, rigorously protested this change.[32] Apparently, President Pérez's own support of commercial farming was an invitation to massive corruption. All previous debts owed to the government were exonerated, massive new projects and funds were directed toward the commercial farm sector, some of the "exonerated" public funds and new loans went directly into land or, more likely, into Miami real estate.[33]

Rene Dumont, a respected leftist and agricultural expert whose work in Cuba earned him the wrath of Fidel Castro, was invited to Venezuela. His 1975 report on Venezuelan agricultural policy was a blistering indictment of populism's failures. Huge amounts of money were spent on new projects such as irrigation systems, while existing facilities nearby remained underutilized or in a state of disrepair. Exonerating existing debts, according to Dumont, punished the hardworking and honest farmers, while it rewarded the dishonest ones.[34]

In 1980, AD sought a scandal to pin on the COPEI administration of sufficient size to balance its Sierra Nevada scandal of the Carlos Andrés Pérez years (1974–79). It had no trouble finding one in the Agricultural Marketing Corp. (CORPOMERCADEO). Not only were the COPEI administrators of CORPOMERCADEO inept and apparently corrupt, but they also violated the basic tenant of the Third-World creed. CORPOMERCADEO had purchased low quality and overpriced corn from the hated regime of South Africa.[35]

In summary, competition for rural electoral support between the two principal populist parties, AD and COPEI, has been sterile. Venezuela, despite adequate land and an abundance of capital, has failed to feed itself. While most of its land is of poor quality, the good land is underutilized.[36] The vast majority of "settled" agrarian-reform peasants live on marginal land and are largely neglected. The true social cost of Venezuela's agrarian failure, however, must include a recognition that from 500,000 to 1 million undocumented Colombians have become Venezuela's true rural, marginal people.

TERRITORIAL POLITICS: COMPETITIVE REGIONAL DEVELOPMENT

Political scientists disagree about the political role of regionalism in a culturally homogeneous nation such as Venezuela, where almost all citizens share the same culture, language, and religion. According to some writers, regionalism is merely a container of political action, caused by other factors. This view overlooks the fact that people who live in the same territory over time encounter the same historical problems and thus develop a common history and identity. This sense of shared space and experience often leads to shared political orientations, regardless of differences in social status or race. (The writer has previously written at length on the enduring strength of the regional issue in Venezuelan politics.[37])

There are two dimensions to regionalism in Venezuela. The first is the center-periphery cleavage: a relatively prosperous, modern, and highly urbanized central-core region surrounding the capital, Caracas, confronts a relatively poor, more traditional, rural periphery. The second contemporary dimension of regionalism is the competitive regionalism of the eastern and western peripheries as each vies with the other for development projects and programs. The politics of regional-territorial development are an ideal arena for nonideological brokerage politics. Competitive regional development issues transcend both the particularistic limitations of patron-clientelism and the stark and disruptive politics of class conflict. All classes and factions of a given community or region unite to support the allocating of a major public investment in their region.[38] Populism could be ideally suited to manage the regional issues.

The origin of this contemporary regional rivalry was the awareness that the raw-material mining sectors of the peripheral regions—the oil of the states of Zulia and Anzoategui and the iron ore of the state of Bolivar—were producing just about all the nation's wealth, while the central region was consuming just about all of it. The accelerating primacy of the central region and metropolitan Caracas was due to the increasing concentration of people, jobs, and money in this core region. Central-government decision making and funding became all-important to the lives of most business leaders and politicians as the state moved to stimulate development. In the words of one regional planner, John Friedmann, the very efforts of the government to sponsor regional development and decentralization attracted more people to Caracas: "A symbiosis between economic and government organizations occurred as politicians, bureaucrats, and businessmen mingle in exclusive clubs and the city's top restaurants, send their children to private schools . . . and form tight social networks of their own."[39]

This quotation illustrates a perennial problem of populism. The elected political leaders of the people tend to mingle and share a life style with the hated oligarchy and foreign representatives. Populist planning for the development of the neglected periphery was to be done in Caracas, untainted by excessive grass-roots, regional input. In 1960, the politics of competitive regional development began with the central-government decision to create the national Venezuelan Guayana Corp. The CVG, whose first director was retired Gen. Alfonzo Ravard, considered itself a national agency designed to develop the iron-ore and hydroelectric resources of the Guayana region (Bolívar state) for the benefit of the entire nation.

The fact that in the 1960s about 10 percent of the total government investment was earmarked for the CVG and the sparsely populated Bolivar state stirred considerable resentment in two important western areas of the nation: the poor but politically important Andes region—the states of Trujillo, Mérida, and Táchira had controlled Venezuelan politics during the Cipriana Castro-Juan Vicente Gómez era—and the oil-rich state of Zulia. These states were also COPEI electoral strongholds and noticeably devout in their Catholic practices. Many in the Andes saw the CVG as an AD plot to permanently marginalize their region's economy. Over the opposition of both the CVG and CORDIPLAN, an Andean political bloc pushed through a national law in 1964 to establish the Venezuelan Andean Corp. (CORPOANDES). CORPOANDES focused its attentions on obtaining public-works projects comparable to those being planned for the Guayana development.

The CORPOANDES law stirred up Zulia state. Zulia and Lake Maracaibo had given Venezuela three-fourths of its oil revenues. Yet in the early 1960s, the state was experiencing a severe economic de-

lombian border) convinced CORPOZULIA and the University of Zulia that Zulia could become like Bolivar state, a steel maker. Unlike the CVG, CORPOZULIA's crusade for a second steel plant, using Bolívar state's iron ore, was strictly a grass-roots effort and, in the words of CORPOZULIA President Fernando Chumaceiro, an example of the failure of the "centralized model of national development." As part of the promising technology-cooperation agreement signed between the United States and Venezuela, Guasare coal has been sent to U.S. Department of Energy laboratories for analysis. A prospective Zulia steel plant was proposed in the 1970s in a nation whose first effort at steel making, CVG-SIDOR, was not an overwhelming success.

CORPOZULIA's concentration on developing a major, steel-based, industrial center on the western side of Lake Maracaibo, adjacent to the city of Maracaibo, stirred some resentment among communities such as Cabimas, Lagunillas, Ciudad Ojeda, and Tia Juana, which had given Venezuela most of its petroleum wealth yet had received only "benign neglect" in return as late as the 1960s. In the 1960s, a group of youthful politicians centered around a local MEP leader, José Bauza González, and COPEI leader Atilio Fossi organized the government of the city of Cabimas (district of Bolivar, Zulia state) into one of the most progressive urban regimes in Venezuela. As the Cabimas group organized its own grass-roots development effort, it also increasingly felt that the eastern-shore communities of Lake Maracaibo were not receiving adequate attention from CORPOZULIA, and it petitioned to form Venezuela's only major, subregional (part of a state), development corporation.

As the municipalities of Cabimas, Cuidad Ojeda, and Lagunillas started to develop plans for their renovation and economic diversification, MARAVEN officials decided to intervene for fear that surface urban projects that ignored the subterranean oil reserves could lead to closing off a significant, future oil production. MARAVEN officials, Cabimas politicians, and the labor and business associations of the region joined together in one of Venezuela's most exciting grassroots development projects: the rescue and renovation of the east coast of Lake Maracaibo.

Two innovative aspects of this ongoing project deserve mention in this study of national politics: Under the direction of the Ministry of Urban Developments (MINDUR), all concerned public agencies formed an interinstitutional working group to tackle the redevelopment in a comprehensive manner,[44] and a functioning, institutionally based, subregional Participatory Assembly was created to ensure adequate grass-roots inputs into the region's redevelopment. Two quite exciting regional business associations, the Association of Merchants and Industrialists of Lagunillas District and the Association of Merchants and Industrialists of Bolivar District, joined with the local labor confederation and the Zulian congressional bloc to ensure that grass-roots participation would be a meaningful exercise. As stated in Chapter 8, the rescue of Lake Maracaibo is proceeding as an exemplary case of true democratic planning.

In a period of economic slowdown, the SIDOR-Zulia project has been scaled down, while the importance of guaranteeing a prolonged economic life to Lake Maracaibo oil has been given top priority.[45] CORPOZULIA has become a key actor in redesigning the east coast of the lake and in linking both sides of the state. The decision to decentralize the headquarters of Petroleum of Venezuela (PDVSA) affiliates means that the number of top executive personnel living in Zulia will increase rapidly in the immediate future. This will lead to demands for improved education, cultural facilities, and the quality of basic public services.

In addition to the healthy competition among the Four Sisters of the PDVSA, the potential regionalization of the affiliates of the PDVSA will firmly link a given affiliate to a region: MENEVEN and LAGOVEN with the Orinoco Valley and MARAVEN and PEQUIVEN with Lake Maracaibo. This development could alter the spatial organization of political decision making in Venezuela. Starting in 1979, local municipal governments have been elected in separate (off-year) elections. While impersonal party ballots were used in 1979, there is now the possibility that political careers on the local level can be advanced by service to one's community, not only to one's party.

The development of the Orinoco Heavy-Oil Belt is being linked to CVG industrial development and the revitalization of agriculture in eastern Venezuela. This linkage is part of the current drive to deepen the meaning of the 1976 nationalization and to dominate oil.[46] The Orinoco Belt is being studied as part of a broader, eastern regional development. The Central University Center for the Study of Development produced a dramatic comprehensive plan for the linked development of the Orinoco Valley,[47] although the Orinoco development was slowed in early 1983.

The fact that metropolitan Caracas indeed became an urban disaster in the 1980s, saved (in the words of respected essayist Juan Liscano) only by its excellent restaurants, unfit for residency by the rich, influential, and powerful, aided the new sense of urgency in regionalizing Venezuela.[48] Liscano and others believe that the uncontrolled growth of Caracas has been a direct result of the failure of populism, the failure of agrarian reform, and the failure of the political parties to be more than self-serving patronage machines. While some are calling for a decision to relocate the capital from Caracas to the "virgin" and true Venezuela south of the Orinoco River—the communities of Caucara and Cabruta—others, citing the failure of Brasilia, prefer not to try another massive "pharaonic" project to solve Caracas's urban congestion.

The growth of grass-roots regional-development agencies and a system of regional cities with adequate schools and other public services and the continued workings of democracy should slowly reduce the territorial and political imbalance in Venezuela.

NOTES

1. Claudio Véliz, ed., *Obstacles to Change in Latin America* (New York: Oxford University Press, 1967), p. 5.

2. A "Panoramic Vision" of the six national plans that cover 1958 to 1985 was included as part of the *VI National Plan 1981–1985—Growth for All*, promulgated by the Hererra government. See Venezuelan Central Office of Planning and Coordination, *VI Plan de la Nacion 1981–1985 Crecimiento para Todos* (Caracas: Central Office of Planning and Coordination, 1981).

3. Fernando Coronil and Julie Skurski, "Reproducing Dependency: Auto Industry Policy and Petrodollar Circulation in Venezuela," *International Organization* 36 (Winter 1982): 70–71.

4. Venezuelan Central Office of Planning and Coordination, *Elementos para una Política de Desarrollo de la Industria Automotriz* 1 (April 1962).

5. The author has published several accounts of the politics of industrial planning in Venezuela. See David E. Blank, *Politics in Venezuela* (Boston: Little, Brown, 1973), pp. 259–66, "Political Conflict and Industrial Planning in Venezuela," in *Venezuela 1969: Analysis of Progress*, ed. Philip B. Taylor, Jr. (Houston: University of Houston Office of International Affairs, 1971), pp. 84–106, "Policy-making Style and Political Development: The Introduction of a System of Democratic Planning in Venezuela, 1958–1968," (Ph.D. dissertation, Columbia University, 1969).

6. Robert N. Gwynne, "Auto Industry—The Scramble Continues: Who Wants What and Where," *Business Venezuela* (November/December 1979): 28.

7. Ibid., p. 29.

8. Coronil and Skurski, "Reproducing Dependency," pp. 69, 76.

9. Gwynne, "Auto Industry," p. 29, and Coronil and Skurski, "Reproducing Dependency," p. 85.

10. Coronil and Skurski, "Reproducing Dependency," p. 82.

11. Gwynne, "Auto Industry," p. 28.

12. Blank, "Political Conflict and Industrial Planning," pp. 99–100.

13. "Economy: Latin Car Production Cranks Up to World's Top," *Latin American Times* 3 (March 1981): 18.

14. Fred Jongkind, "Industrialists and the State: Private Views on the Public Role in the Industrialization of Venezuela," (Paper presented to the Latin American Studies Association, Pittsburgh, April 1979).

15. Ibid.

16. The best assault on the insidious myth of marginality is Janice Perlman, *The Myth of Marginality: Urban Poverty and Politics in Rio de Janeiro* (Berkeley: University of California Press, 1978).

17. Normal Gall, "Oil and Democracy in Venezuela, Part II: The Marginal Man," *American Universities Field Staff Reports* 7 (January 1973): 8, and Leda Siqueira Wiarda, *Family Planning Activities in a Democratic Context: The Case of Venezuela* (Private printing and copyright, 1979).

18. Steve Ellner, *Los Partidos Políticos y la Disputa por el Control del Movimiento Sindical en Venezuela, 1936–1948* (Caracas: Universidad Catolica Andrés Bello, 1980).

19. Bernard Lestienne, *El Sindicalismo Venezolano* (Caracas: Centro Gumilla, 1981), p. 24.

20. José Ignacio Arrieta, "¿Porque Intervinieron a SUTISS?," *SIC* (December 1981), "La Utonomia Sindical esta en Peligro," *El Nacional*, December 21, 1981, "Como Intervenir un Sindicato," *Resúmen* December 6, 1981.

21. "Venezuela's Unions Up for Grabs," *Latin American Weekly Report*, March 19, 1982, p. 8.

22. J. Raúl Alegrett, "The Venezuelan Agrarian Reform: Impact and Perspectives," in *Venezuela: 1969 Analysis of Progress*, ed. Phillip B. Taylor, Jr. (Houston: University of Houston Office of International Affairs, 1971), p. 75.

23. Ibid.

24. Judith Ewell, *Indictment of a Dictator: The Extradition and Trial of Marcos Pérez Jiménez* (College Station, Texas: A. & M. University Press, 1981), p. 54.

25. George W. Schuyler, *Hunger in a Land of Plenty* (Cambridge, Mass.: Schenkman, 1980), p. 160.

26. Ibid., p. 162.

27. Bill Stewart, *Change and Bureaucracy: Public Administration in Venezuela* (Chapel Hill: University of North Carolina Press, 1978), pp. 22–23. Stewart draws this conclusion on the works of John Duncan Powell, *Political Mobilization of the Venezuelan Peasant* (Cambridge, Mass.: Harvard University Press, 1971), and John Kirby, "Venezuela's Land Reform," *Journal of Inter-American Studies and World Affairs* 15 (May 1973).

28. Robert P. Matthews, presentation to the Latin American Studies Association, Washington, D.C., March 1982, and Schuyler, *Hunger in a Land*, p. 58.

29. Schuyler, *Hunger in a Land*, p. 185.

30. "Los Grandes Temas, Reforma Agraria: Balance 20 Años de un Desafio," *El Diario de Caracas*, August 10, 1980, p. 10.

31. Schuyler, *Hunger in a Land*, pp. 66–67.
32. Ibid., p. 151.
33. Howard Handelman, "Scarcity Amidst Plenty in Oil Rich Venezuela," *Common Ground* 4, Spring 1978, p. 52.
34. Ibid.
35. "Maiz Surafricano," *Plural* (May 1980).
36. Handelman, "Scarcity Amidst Plenty," p. 53.
37. David E. Blank, "The Regional Dimension of Venezuelan Politics," in *Venezuela at the Polls*, ed. Howard R. Penniman (Washington, D.C.: American Enterprise Institute, 1980).
38. Ibid. See the author's use of the contribution of Arturo Valenzuela to our understanding of brokerage politics, *Political Brokers in Chile: Local Government in a Centralized Policy*, p. 199.
39. John Friedmann, "The Spacial Organization of Power in the Development of Urban Systems," in *Regional Policy: Readings in Theory and Application*, ed. J. Friedmann and William Alonso (Cambridge: Massachusetts Institute of Technology Press, 1975), pp. 272–73.
40. See the 1963 report on the economic emergency in Zulia prepared by the state FEDECAMARAS body with the help of the CTV, University of Zulia, and Gov. Luis Vera Gómez: *El Zulia en Estado de Emergencía Economica* (Maracaibo: Camara de Comercio de Maracaibo, 1963).
41. For a detailed description of the El Tablazo, Tuy Medio, and Ciudad Guayana projects, see Alan Turner and Jonathan Smuliani, "New Cities in Venezuela" *Town Planning Review* (Liverpool) 42 (1971), Lloyd Rodwin, ed., *Planning Urban Growth and Regional Development: The Experience of the Guayana Program of Venezuela* (Cambridge: Massachusetts Institute of Technology Press, 1969), Fernando Travieso, *Ciudad, Region y Subdesarrollo* (Caracas: Fondo Editorial Comun, 1972), and David E. Blank, "The Great Leap Spaceward: The Politics of Planning New Cities in Mexico and Venezuela," (Paper read at the national meeting of the Latin American Studies Association, Pittsburgh, April 1979).
42. Luis Lander and Cruz Fernández, *Informe: Comision para el Desarrollo Urbano y la Vivienda* (Report to the President on Urban Development and Housing, Caracas: December 11, 1964).
43. David J. Myers, "Urban Voting, Structural Cleavages, and Party System Evolution: The Case of Venezuela," *Comparative Politics* 8 (October 1975), pp. 119–51, and David J. Myers, "Caracas: The Politics of Intensifying Primacy," *Latin American Urban Research*, vol. 6 (Beverly Hills, Calif.: Sage, 1978), pp. 227–58.
44. *Sintesis del Trabajo Realizado por el Grupo Interinstitutional de la Costa Oriental del Lago de Maracaibo* formed in 1979 comprised of MINDUR, CORDIPLAN, CONZUPLAN, PDVSA, and CORPOZULIA representatives.
45. Fernando Chumaceiro, *El Desarrollo de la Costa Oriental del Lago en el Contexto del Desarrollo Regional* (Ligunillas: Association of Merchants and Industrialists of Ligunillas: 1979, available at the office of CORPOZULIA, Maracaibo).
46. COPEI Vice-minister of Energy José Ignacio Moreno León has become a strong advocate of intimately linking oil to agricultural and industrial development, "Con la Explotación de la Faja—El Petróleo Impulsara el Desarrollo Industrial," *El Nacional*, July 19, 1981. See his *Profundización de la Nacionalización Petrolera Venezuolana* (Caracas: Ediciones Centauro, 1981).

47. Venezuelan Central University Center for the Study of Development, *Estudio Prediagnostico para el Plan Maestro de Ordenamiento Territorial del Area de la Faja Petrolifera del Orinoco* (Caracas: Central University Center for the Study of Development, 1979).

48. Juan Liscano, "Caracas esta condenada," in *Caracas, La Horrible*, ed. Luis Buitrago Segura (Caracas: Ateneo de Caracas, 1980).

10

THE FUTURE OF DEMOCRACY

Venezuela faces a challenging decade of diminished oil revenues and an acknowledged disenchantment with the workings of its populist democracy. The sixth post-1968 general elections took place in 1983 amid growing unemployment and business bankruptcy. The viability of electoral democracy in Venezuela has been challenged by citizens and foreign journalists alike. A series of critical articles in the New York *Times* in March 1983 featured reports of governmental paralysis and a growing popular disaffection from populist politics.[1] One of these articles quoted a middle-class farmer, frustrated with government mismanagement and corruption regarding agriculture as well as with his own problems in obtaining water and credits: "Pérez Jiménez built those . . . irrigation canals. . . . The democratic government ruined them."[2]

In response to the implication that populist democracy wasted 25 years and that the Pérez Jiménez dictatorship was more productive, Venezuela's ambassador to Washington, Dr. Marcial Pérez-Chiriboga, wrote a letter to the editor of the New York *Times*, asserting that "Democratic Venezuela is coping with trying times."[3] Words such as "austerity" and "retrenchment" dominated the 1983 election campaign. There are those in Venezuela who see the present economic disaster as a necessary purgative or tonic in order to move beyond the limits of populism. As previously stated, the windfall profits of the petrobonanza years caused more harm than good in developing Venezuela into a more just, participatory, and autonomous nation. The present crisis does appear to be engendering creative policies and a new determination that democracy will not only survive but also more purposively attack the problems of dependency and social injustice.

There are several reasons for this optimism. Although the currently weak oil market and Venezuela's increasing dependence on less profitable heavy oil is discouraging, the long-term prospects for oil should provide Venezuela with sufficient resources to avoid the economic catastrophe that would doom democracy. Venezuela's admittedly exasperating electoral populism angers and disillusions many, but it avoids threatening crucial power elites to the point of their moving against it. There may be no popular mandate for radical change in Venezuela. Finally, this author's optimism is, in part, based on the realization known to critical Venezuelans that the authoritarian alternative has itself been tarnished.

Using Brazil as a model, in early 1983, that nation's international debt was in the $80 billion to $90 billion range. This debt was partly engendered by the big-power megalomania of its technocratic authoritarian leadership in the 1964–82 period and partly by the disincentives given Brazil's poor during the period of the Brazilian "miracle." In the fall of 1982, Brazil held its first competitive legislative and state elections in more than 15 years, and Brazil itself is evolving toward the complex mixture of electoral populism, planning, and praetorianism characterized by Venezuela. The 1980s in Latin America have been marked by the "liberalization" of authoritarian regimes, such as that of Brazil, rather than by the atrophy of existing electoral democracies.

The task and dilemma facing Venezuela's political sector is how to assume a more purposive leadership of the nation without appearing to such critical groups as the military and business communities that it has become radicalized and a threat to their position in the nation. In reviewing the decline and fall of democratic populism in the 1960s and 1970s, Robert Wesson developed the following scenario.[4] According to Wesson, populism in Latin America has been the adaptation of traditional cultural norms of clientelism to a mass electoral context. Patronage and reform rhetoric suffice in times of social expansion and economic prosperity.

In times of economic and social deterioration, according to the Wesson scenario, the populists panic and become increasingly radicalized. They blame the "imperialists of the International Monetary Fund" for their budgetary excesses, the national oligarchs and business community for their own erstwhile policy timidity, and "carried away by their own rhetoric . . . believe themselves stronger than they are; challenge the military. . . but the populist government is toppled almost as easily as the oligarchic regimes of the past."[5]

On one hand, Venezuela's political establishment recognizes the limitations of populism, while on the other hand, it understands the need for prudence and, above all, the need not to rock the boat by generating excessive social conflict. Eduardo Fernández, Social Christian Party secretary general, suggested in the 1980–81 period that populism was a normal, "infantile" stage of democracy. One of Democratic Action's top leaders, Luis Esteben Rey, is quoted as saying that the populist style can be compared to "trying to put out a fire by dousing it with [more] money." This style of problem solving was limited even in times of a petrobonanza. U.S. readers should appreciate that the coming of mass electoral democracy during the age of Jackson (1824–30) was also the emergence of the spoils system and the demise of erudite policy discussions in their country.

It is one thing for political leaders to come to a consensus concerning what needs to be done and it is quite another matter for them to implement the hard decisions necessary to translate good intentions into effective public policies. There are some indications, however, that the political sector is working to fulfill its responsibilities.

The most remarkable feature of the 1983 electoral campaign in Venezuela was its lack of alarming drama. All eight candidates insisted on the need for a change and all agreed that a more purposive democratic manner was needed. One might assume that the discontent and the disillusionment reported in this book are finding an outlet in electoral process.

The fact that neither the left nor the right could, as of June 1983, unify behind a single credible candidate to challenge the electoral hegemony of the Democratic Action (AD) and the Social Christian Party (COPEI) (the parties of the post-1958 establishment), is also a sign of the system's stability. A truly threatening political crisis is often anticipated by a polarization of the vote on the extremes and a diminution of the vote of the center. The polls suggested that when the December 1983 votes were counted, this would not be the case. AD and COPEI were expected to maintain their hold on 70 to 80 percent of the vote.

Rafael Caldera, COPEI's 1983 candidate, was president of Venezuela in the 1969–74 period. He took the candidacy away from the choice of the incumbent COPEI president. Caldera is a son of the western plains states and, in regional brokerage terms, is also the candidate of the development claims of Zulia, the Andes, and western plains regions as well as that of COPEI.

Jaime Lusinchi won AD's nomination as the candidate of its Labor Bureau and of the Bétancourt faction. Few would, however, state that former AD president, Carlos Andrés Pérez, was definitely eclipsed. Lusinchi was born in the eastern state of Sucre and, in regional brokerage terms, he is the candidate of the eastern Guayana and Orinoco Belt regions. Would a Lusinchi administration allow the scaling down now underway of the Orinoco Oil Belt development or the end to the revitalization of the Guayana industrial complex?

Lusinchi, heavily favored to win the December 1983 presidential election as of mid-1983, would come to that office with a unique party background. He has never held an executive position on the national level and has seldom traveled outside of Venezuela.

Neither of the two competitive candidates of the Left—Teodoro Petkoff and José Vicente Rangel—posed a threat to the AD-COPEI electoral hegemony. Petkoff was once a Communist guerrilla leader, but he is now an irreconcilable anti-Soviet Eurocommunist. Rangel, who was never a Communist, is now saying nice things about Fidel Castro, but in the context of Venezuela's present attempt to move into the nonaligned movement, Rangel is not all that radical. There is much of a personal feud in the Petkoff-Rangel clash.

They agree only on one issue, that a leftist-unity candidate would have enhanced the left's electoral credibility. The 1983 general election was for them a primary election for control of the left. The innovative radicals in the Radical cause (*Causa R*) stayed aloof from the Petkoff-Rangel feud and suggested independent editor Jorge Olivarría as a third leftist-unity candidate.

Jorge Olivarría and a second journalist, Rafael Poleo of *Zeta*, were briefly jailed by President Herrera Campins for "insulting" the Office of the Presidency and the memory of liberator Simon Bolívar. This outrage against the freedom of expression was due to the humorous way in which Venezuela's current economic crisis was portrayed in their magazines. This jailing for a 15-day sentence engendered much public criticism and a political scandal that may have hurt COPEI's electoral chances. While Venezuela's public has lost its patience with many of its populist politicians, it has reaffirmed its democratic learnings and, equally important, its sense of humor.

Retired Gen. Rangel Bourgoin insists that a National Rescue is needed, while retired Gen. Castro Hurtado prefers a New Order. One should not automatically label Castro Hurtado the extreme right. He was willing to team up with former guerrilla leader, Douglas Bravo. This electoral maneuver earned him a reputation for being "eccentric" and apparently lost him the support of the conservative civilians in the New Generation Movement. New Generation had to decide on one of its own as its presidential candidate.[6]

DEMOCRACY AND DEPENDENCY

Oil is critical to Venezuela, reflected in the title of this book, *Politics in a Petroleum Republic*. The question concerning democracy's ability to lessen national dependency in the critical oil sector is best rephrased as whether populist democracy can eventually convert Venezuela into an economically and socially integrated nation of prosperous farms and productive factories with oil as one of its principal exports or whether democracy is destined to provide human rights and patronage as part of an electoral sideshow to an oil republic.

The soundness of the Venezuelan decision to nationalize oil with a minimum of conflict with the transnational companies was demonstrated by the fact that both production and sales maintained themselves in the critical six-year period after nationalization. This achievement was accomplished after a decade of disinvestment in the industry. The initial "somewhat inflated technical service fees" given the companies in 1976 apparently were the price Venezuela played for this production stability and transnational cooperation.[7] However, these contracts were substantially renegotiated in 1979, and in addition to reducing the fees, the hated confidentiality clause, which prevented one PDVSA affiliate from sharing its know-how with its sisters, was eliminated.[8] Much of the charge that the 1976 nationalization was fraudulent was based on the existence of this confidentiality clause.

This writer argues that despite its many failings, populist democracy is, indeed, winning the war against dependency. A more politically accountable but not politicized (in the patronage sense) management of oil is being achieved. The elected government, not the transnational oil corporations or the managers of Petroleum of Venezuela S.A. (PDVSA), is now establishing the framework for the industry's future. The policies of globality—maintaining Oil Producing and Exporting Countries (OPEC) and diversifying its markets, deepening the meaning of nationalization (by imposed national procurement of PDVSA needs), government-to-government arrangements for exchanging oil for technology, and regionalization (making PDVSA operations affiliates more accountable to their grass roots)—are all binding oil to the nation.

This is a vastly different situation than that portrayed in a 1979 article that asserted that the oil nationalization merely "redefined the structure of dependency" and "legitimized imperalism."[9] The author of the 1979 article, Vegard Bye, listed seven indicators of the structure of dependency. From the perspective of late 1983, this author suggests that the ill effects of these indicators, their distortion of Venezuelan society, and their disruption of national integrity have all lessened. To sum up the writer's evaluation of the lessening of dependency, I shall use Bye's seven indicators.

> Economic Penetration. The increased use of government-to-government, long-term sales contracts and the imposition of national procurement on the PDVSA has greatly lessened this. The intimacy between the PDVSA affiliates and their former prenationalization alter egos, Exxon and Shell, has been broken.
>
> Technological Penetration. The ability of both the Venezuelan and the Federal Republic of Germany governments to develop government-to-government programs of technology cooperation and the ability of PDVSA's technology-research institute to coproduce a suitable technology of heavy-oil upgrading with the German firm VEBA are outstanding examples of Venezuela's new oil autonomy. This is the first time since nationalization that the PDVSA has formally associated with a foreign firm. Congressional scrutiny of the PDVSA-VEBA partnership is going to be close. In achieving the PDVSA-VEBA deal, Venezuela rejected the alternative technology offered by the powerful Bechtel Engineering Group and a second U.S. firm. The Bechtel process of upgrading heavy oil was ill-suited to Venezuela due to its costs and the need to produce tons of disposable carbon waste.
>
> Trade-Partner Concentration. An aggressive sales policy, Venezuela's tradition of reliability, intra-OPEC trades, and the policy of globality meant that about 50 percent of Venezuelan oil in 1982 was sold to new customers such as Brazil, the Federal Republic of Germany, and Japan.
>
> Dependence on Foreign Trade. Venezuela can no longer afford to import 60 to 70 percent of its food. This is quite an incentive to domestic farmers and to a revitalized peasant sector.

Monoproduction. The devaluation of the bolivar has freed the national currency from oil. In the long run, this means that Venezuelan goods will be more competitive in the Andean Pact market and that Venezuela's proclivity to import is lessened.

External Exploitation. This is an emotional concept. The ability of Venezuela to negotiate with the United States, France, Canada, and the Federal Republic of Germany before the breakthrough accord with Germany indicates that it has gained an ability to deal on terms of equality. The rejection of Bechtel as the master planner of the Orinoco heavy-oil development and the willingness to scale down the development are both indicative of Venezuela's new autonomy.

Enclave Structure. The development of comprehensive interministerial groups for the rescue of Lake Maracaibo, the scaled-down Orinoco development, and the new grass-roots political clout of the municipalities should end the legacy of the enclave. Gone are the days when oil was developed in isolation.

The basic reason for the lessening of dependency is that the enlightened self-interest of a variety of such groups as the College of Engineers and the recession-troubled business community, operating through a pluralist democracy, has tilted the scale against maintaining the PDVSA's autonomy of action. These professional and business groups have joined with more radical elements in the political sector and the universities to make the PDVSA more accountable. This means that via the new national procurement policy, the PDVSA will now buy Venezuelan and hire Venezuelan. The scaling down or redimensioning of the Orinoco development makes its Venezuelanization more feasible. Gone is the automatic reliance on the transnationals, MARAVEN with Shell, LAGOVEN with Exxon, and MENEVEN with Gulf, and their high-tech megaprojects.[10]

There is also evidence to suggest that Venezuela is operating under the assumption that its oil sales will average about 1.5 million barrels per day over the next decade, rather than the present 1.8 million to 1.9 million or the previously projected 2 million. The high cost and long lead time required for the Orinoco development and newer, less expensive developments outside of Venezuela seem to be the prime reasons for this expected decline.[11] This development translates into a lessened reliance on the easy money of the oil sector.

There are five key questions that Venezuela's ongoing democratic populism will address over the next decade. While its response to these questions will be less dramatic than that proposed by its authoritarian critics (either of the radical-revolutionary or conservative-technocratic variety), this writer senses that the Venezuelan response may be more lasting and profound than any authoritarian alternative.

The first question, the most pressing, involves whether a politically accountable but not politicized management of the vital oil industry can be achieved. The honeymoon between the PDVSA and Venezuela is over, and the dominant personality of retired Gen. Alfonzo Ravard should pass from the scene. There is every indication, however, that the basic structures of the democratic management of the industry created in the 1976–82 period will survive. In recognition of the special international role the PDVSA has played, rumor has it that the political parties are in agreement that the PDVSA's present legal counselor, Andrés Aguilar, will be its next president. Aguilar has assumed key diplomatic roles under both Organization of American States and United Nations auspices regarding human rights and the Iranian revolution.

In August 1983, President Hererra Campins announced the retirement of retired General Alfonzo Ravard as President of the PDVSA. Named as his replacement was none other than Energy Minister Humberto Calderón Berti. While Calderón Berti is a respected geologist with experience in the PDVSA as the head of its technical affiliate, INTEVEP, who earned a world-wide reputation for his astute command of petroleum issues as Minister of Energy, his appointment was viewed with alarm by many Venezuelans. Since its creation in 1976, the national consensus has been not to politicize the management of this most vital public enterprise. Despite his professional credentials, Calderón Berti is viewed essentially as a COPEI politician, and his appointment threatens the consensus that the PDVSA must never be allowed to degenerate into a typical inefficient Venezuelan public enterprise. Fear already has been expressed that some of the now scarce PDVSA funds are being diverted to the COPEI election campaign.

The imposition of a system of democratic management over the PDVSA supported by this writer is quite different from its politicalization. Democratic management has placed a close democratic oversight and vigilance on the PDVSA requiring it to better integrate itself within Venezuela's economy and to cut its umbilical cord to the transnational petroleum companies. Between 1981 and 1983, PDVSA's management has responded to the new rules of the game by adhering to the policies of national procurement (of its material and professional needs), regionalization, and integrating itself with governmental and grass roots community development efforts. This "deepening of the meaning" of nationalization has been achieved without lessening the company's operational efficiency at a time of world oil glut. Democratic management respects the need to preserve professionally qualified administrators in charge of the PDVSA and the indispensable role played by merit in running a vital public industry.

There are many indicators that Venezuela's populist democracy will in fact make this feared politicalization of the PDVSA a short interlude. Both AD's presidential candidate, Jaime Lusinchi, and COPEI's Rafael Caldera, have indicated that Calderón Berti will be removed from the PDVSA. The nation's press, business community, and many professional associations have also called for the preservation of the PDVSA free from partisan and patronage control. Some military officers have also expressed their personal irritation at the treatment of the respected, retired General Alfonzo Ravard. There is every indication that, in March 1984, a respected professional such as Andrés Aguilar will be PDVSA's president.

The second question involves the relationship between populism and its present mass clienteles. How can labor union autonomy be secured without disrupting the already frail industrialization effort? This question is especially acute in the public-sector enterprises with their legacy of feather bedding. Increasing productivity and workplace decorum consistent with democracy is difficult under any circumstances. The fact that the reputation of the top leaders of the Central Labor Confederation has been tarnished by the debacle of the Bank of the Venezuelan Workers bankruptcy means that these old political-party warhorses can do little to brake the rising anger of the youthful Venezuelan industrial worker.

The third question involves whether the remaining peasant sector can be truly integrated into national economic life via peasant cooperatives and peasant enterprises or whether the Venezuelan peasant is to remain merely a welfare clientele of the populist political parties? Venezuela can no longer afford to import in excess of 60 percent of its basic food supplies. The expected shortfall cannot be met only by the commercial farms sector. Juan Liscano, a former speech writer for the late President Betancourt, considered the transformation of the once-sturdy Venezuelan peasant (*campesino*) into a phlegmatic urban consumer living in a slum with a contraband color television set to be the greatest failure of populist democracy.

Caracas in the early 1980s should have reached a population of approximately 4 million, or 25 percent to 30 percent of the nation's total. Whether a limit to the growth of the urban disaster can be reached before the congestion, contagion, contamination, and crime overwhelm Venezuela may be as critical as the question relating to oil. The capacity of Venezuela to provide basic necessities such as drinking water and a steady flow of electricity to Caracas is about to be exhausted. Sewer treatment and garbage disposal is already a disaster. There is no room left for necessary new highways, school sites, and playgrounds. Liscano has also blamed the neurosis of a weak national identity "on the uncontrolled growth of Caracas which in turn reflected the failure of agrarian reform" and industrialization.[12]

While there has been much criticism expressed in this volume about the particularistic and personalistic cultural foundations of populist democracy, Venezuela's cosmopolitan upper class and worldly managerial elite should not escape their share of the blame. It is this writer's opinion that the entrepreneur's justification is his willingness to take risks with his own money in order to introduce new products and improvements. Venezuela's upper class is known for its manipulation and speculative abilities in developing the previously mentioned marriage of complicity with the top populist politicians rather than for its risk-taking abilities. If the upper-class Venezuelan would invest as much in his country as he has in Miami, Venezuela would have many more productive farms and factories. Venezuela's upper class also has one of the world's lightest tax burdens.

In conclusion, what kind of democracy will Venezuela have in the rest of the 1980s and into the 1990s? The prospects are that it will be more of the same, one hopes with a greater sense of purpose and a somewhat-diminished reliance on patronage and paternalism. Charles E. Lindblom, who wrote of "The Science of Muddling Through" in 1959, wrote a defensive article as part of a symposium on pluralist policy making in 1979. The title of the 1979 article was appropriate to contemporary Venezuela: "Still Muddling, Not Yet Through."[13]

> Incrementalism in politics is not in principle, slow moving. It is not necessarily therefore a tactic of conservatism. A fast-moving sequence of small changes can more speedily accomplish a drastic al-

> ternation of the status quo than can an only infrequent major policy change. . . . Incremental steps can be made quickly because they are only incremental. They do not rock that boat, do not stir up the great antagonisms and paralyzing schisms as do proposals for more dramatic change.[14]

A paraphrase of the title of Lindblom's 1979 article describes Venezuelan democracy in 1983: Not Yet Through, Still Muddling.

NOTES

1. New York *Times*, March 20, March 21, March 23, and March 26, 1983.
2. "For Venezuelans, Worries are Replacing Revelry," New York *Times*, March 20, 1983.
3. Letter to the editor from Venezuelan Ambassador Marcial Pérez-Chiriboga, New York *Times*, March 31, 1983.
4. Robert Wesson, senior research fellow, Hoover Institution, "Populism and Military Coups," (Paper read at a conference on "The New Military Politics in Latin America," Hoover Institution, Stanford University, Stanford, Calif., August 27–29, 1981.
5. Wesson, "Populism," p. 31.
6. "Quarterly Economic Review: Venezuela," *The Economist* (3rd quarter, 1982): 10.
7. George Philip, *Oil and Politics in Latin America: Nationalist Movements and State Companies*. (Cambridge, Mass.: Cambridge University Press, 1982), p. 475.
8. Philip, *Oil and Politics*, p. 478.
9. Vegard Bye, "Nationalization of Oil in Venezuela: Redefined Dependence and Legitimation of Imperialism." *Journal of Peace Research* (1979), pp. 57–78.
10. The Federal Republic of Germany Social Democratic Party founded the Friedrich Ebert Foundation social-science research arm in Caracas, the Latin American Institute for Social Research. The institute organized a 1979 roundtable on "Petroleum and Ecodevelopment in Venezuelan," ostensibly to honor the memory of Juan Pablo Pérez Alfonzo. The fiscal and administrative autonomy exercised by the PDVSA and its cooperative relations with the transnationals were severely attacked by participants at the conference and in a posthumously published article by Pérez Alfonzo. See Dorathea Mezger, ed., *Petroleo y Ecodesarollo en Venezuela* (Caracas: Latin American Institute for Social Research, Ateneo de Caracas, 1981).
11. This forecast was made by the respected London Intelligence Unit of *The Economist*, "OPEC and the World Oil Outlook: Rebound of the Exporters," *The Economist* (February 1983), mentioned in *Latin American Regional Report: Andean Group* (March 1983).
12. Juan Liscano, "Caracas esta Condenada" in *Caracas, La Horrible*, ed. Luis Buitrago Segura (Caracas: Ateneo de Caracas, 1980), p. 28.

13. Charles E. Lindblom, "Still Muddling, Not Yet Through," *Public Administration Review* 39 (November/December 1979).

14. Ibid., p. 520.

BIBLIOGRAPHY

Alexander, Robert J. *Rómulo Betancourt and the Transformation of Venezuela*. Rutgers, N.J.: Transaction Books, 1982.

———. *The Venezuelan Democratic Revolution: A Profile of the Regime of Rómulo Betancourt*. New Brunswick, N.J.: Rutgers University Press, 1964.

Anderson, Charles W. *Politics and Economic Change in Latin America: The Governing of Restless Nations*. Princeton: Van Nostrand, 1967.

Baloyra, Enrique A., and John D. Martz. *Political Attitudes in Venezuela: Societal Cleavages and Public Opinion*. Austin: University of Texas Press, 1979.

Betancourt, Rómulo. *Venezuela's Oil*. London: George Allen and Unwin, 1978.

Blank, David E. "The Great Leap Spaceward: The Politics of Planning New Cities in Mexico and Venezuela." Paper delivered at the meeting of the Latin American Studies Association, Pittsburgh, April 1979.

———. *Politics in Venezuela*. Boston: Little, Brown, 1973.

———. "Oil and Democracy in Venezuela" *Current History* 78 (February 1980).

———. "Political Conflict and Industrial Planning in Venezuela." In *Venezuela: 1969 Analysis of Progress*, edited by Philip B. Taylor, Jr. Houston: Office of International Affairs, University of Houston, 1971.

———. "The Regional Dimension of Venezuelan Politics." In *Venezuela at the Polls: The National Elections of 1978*, edited by Howard R. Penniman. Washington, D.C.: American Enterprise Institute, 1980.

Bigler, Gene. "The Decentralized Public Administration and Venezuelan Political Economic Development." Ph.D. dissertation, Johns Hopkins University, 1978.

———. "Political Restraint and Military Professionalization in Venezuela." Paper delivered at the conference on the New Military Politics in Latin America, Hoover Institution, Stanford, Calif., August 1981.

Bond, Robert D. *Contemporary Venezuela and Its Role in International Affairs*. A Council on Foreign Relations Book. New York: New York University Press, 1977.

Chalmers, Douglas A. "Parties and Society in Latin America." Paper delivered at the American Political Science Association annual meeting, Washington, D.C., September 1968.

Coronil, Fernando, and Julie Skurski. "Reproducing Dependency: Auto Industry Policy and Petrodollar Circulation in Venezuela." *International Organization* 36 (Winter 1982).

Di Tella, Torcuato. "Populism and Reform in Latin America." In *Obstacles to Change in Latin America*, edited by Claudio Véliz. London: Cambridge University Press, 1965.

Ellner, Steven. "Political Party Dynamics in Venezuela and the Outbreak of Guerrilla Warfare." *Inter-American Economic Affairs* 34 (Autumn 1980).

———. "The Venezuelan Political Party System and Its Influence on Economic Decision Making on the Local Level." *Inter-American Economic Affairs* 36 (Winter 1982).

Ewell, Judith. "The Development of Geopolitical Thinking in Twentieth Century Venezuela." Paper delivered at the meeting of the Latin American Studies Association, Bloomington, Ind., October 1980.

———. *Indictment of a Dictator: The Extradition and Trial of Marcos Pérez Jiménez*. College Station, Texas: A. & M. University Press, 1981.

Friedmann, John. "The Spatial Organization of Power in the Development of Urban Systems." In *Regional Policy: Readings in Theory and Application*, edited by John Friedmann and William Alonso. Cambridge: Massachusetts Institute of Technology Press, 1975.

Fuad, Kim. "Oil Operators: The Forgotten Foursome." *Business Venezuela* (November/December 1981).

———. "Venezuela's Role in OPEC: Past, Present and Future." In *Contemporary Venezuela and Its Role in International Affairs*, edited by Robert D. Bond. A Council on Foreign Relations Book. New York: New York University Press, 1977.

Garcia Araujo, Mauricio. "The Impact of Petrodollars on the Economy and Public Sector of Venezuela." Paper delivered at the meeting of the Latin American Studies Association, Washington, D.C., March 1982.

Gil Yepes, José Antonio. "Entrepreneurs and Regime Consolidation." In *Venezuela: The Democratic Experience*, edited by John D. Martz and David Myers. New York: Praeger, 1977.

Hirschman, Albert O. *Journeys Toward Progress: Studies of Economic Policy-Making in Latin America*. New York: The Twentieth Century Fund, 1963.

Hennessy, Alistair. "Latin America." In *Populism: Its Meaning and National Characteristics*, edited by Ghita Ionescu and Ernest Gellner. London: Macmillian, 1969.

Herman, Donald L. *Christian Democracy in Venezuela*. Chapel Hill: University of North Carolina Press, 1980.

Jongkind, Fred. "Industrialists and the State: Private Views on the Public Role in the Industrialization of Venezuela." Paper delivered at the national meeting of the Latin American Studies Association, Pittsburgh, April 1979.

Levine, Daniel H. *Conflict and Political Change in Venezuela*. Princeton, N.J.: Princeton University Press, 1973.

Lieuwen, Edwin. *Petroleum in Venezuela: A History*. New York: Russell and Russell, 1959.

Mann, Joseph Jr. "Merhav's Venezuela: Poor Little Rich Country." *Business Venezuela* (July/August 1980).

Martz, John D. *Acción Democrática: Evolution of a Modern Political Party in Venezuela*. Princeton, N.J.: Princeton University Press, 1966.

———. "Policy-Making and the Quest for Consensus: Nationalizing Venezuelan Petroleum." *Journal of Inter-American Studies and World Affairs* 19 (November 1977).

Martz, John D. and David J. Myers, eds. *Venezuela: The Democratic Experience*. New York: Praeger, 1977.

Myers, David J. "The Politics of Intensifying Primacy." In *Latin American Urban Research*. Vol. 6. Beverly Hills, Calif.: Sage, 1978.

———. "Urban Voting, Structural Cleavages and Party System Evolution: The Case of Venezuela." *Comparative Politics* 8 (October 1975).

———. "Venezuela's MAS." *Problems of Communism*. (September/October 1980).

Nelson, Joan M. *Access to Power: Politics and the Urban Poor in Developing Nations*. Princeton, N.J.: Princeton University Press, 1979.

Newton, Ronald C. "Latin American Populism: Some Notes on Periodization." In *The Continuing Struggle for Democracy in Latin America*, edited by Howard J. Wiarda. Boulder, Colo.: Westview Press, 1980.

Pérez Sainz, Juan Pablo, and Paul Zarembka. "Accumulation and the State in Venezuelan Industrialization." *Latin American Perspectives* 6 (Summer 1979).

Penniman, Howard R., ed. *Venezuela at the Polls: The National Elections of 1978*. Washington, D.C.: American Enterprise Institute, 1980.

Petras, James, ed. *The Nationalization of Venezuelan Oil*. New York: Praeger, 1977.

———. "Revolution and Guerrilla Movements in Latin America: Venezuela, Guatemala, Columbia and Peru." In *Latin America: Reform or Revolution?*, edited by James Petras and Maurice Zeitlin. Greenwich: Fawcett, 1968.

Peattie, Lisa R. *The View From the Barrio*. Ann Arbor: University of Michigan Press, 1968.

Philip, George. *Oil and Politics in Latin America: Nationalist Movements and State Companies*. Cambridge, Mass.: Cambridge University Press, 1982.

Powell, John D. "Peasant Society and Clientelist Politics." *American Political Science Review* 64 (June 1970).

Rabe, Stephen G. *The Road to OPEC: United States Relations with Venezuela 1919–1976.* Austin: University of Texas Press, 1982.

Rodwin, Lloyd, ed. *Planning Urban Growth and Regional Development: The Experience of the Guayana Program of Venezuela.* Cambridge: Massachusetts Institute of Technology Press, 1969.

Schuyler, George W. *Hunger in a Land of Plenty.* Cambridge, Mass.: Schenkman, 1980.

Silva Michelena, José A. *The Illusion of Democracy in Dependent Nations.* Cambridge: Massachusetts Institute of Technology Press, 1971.

Sosa Rodríquez, Julio. "U.S. and Venezuela: The Oil Diplomacy." *Latin American Times* 3 (March 1981).

Taylor, Philip B. Jr. *The Venezuelan Golpe de Estado of 1958.* Washington, D.C.: Institute for the Comparative Study of Political Systems, 1968.

Tugwell, Franklin. *The Politics of Oil in Venezuela.* Stanford, Calif.: Stanford University Press, 1975.

———. "Review Essay: Energy and Political Economy." *Comparative Politics* 13 (October 1980).

———. "Venezuela's Oil Nationalization, The Politics of Aftermath." In *Contemporary Venezuela and Its Role in International Affairs*, edited by Robert D. Bond. A Council on Foreign Relations Book. New York: New York University Press, 1977.

Vallenilla, Luis. *Oil: The Making of a New Economic Order—Venezuelan Oil and OPEC.* New York: McGraw-Hill, 1975.

Valenzuela, J. Samuel, and Arturo Valenzuela. "Review Essay: Modernization and Dependency: Alternative Perspectives in the Study of Latin American Underdevelopment." *Comparative Politics* 10 (July 1978).

Villela, Annibal A., and Werner Baer. "Evaluating Public Enterprises: Some Dilemmas in Formulating Guidelines." Paper delivered at the first international seminar on Public Enterprises and Economic Development, Brasilia, July 1981.

Wynia, Gary W. *The Politics of Latin American Development*. Cambridge, Mass.: Cambridge University Press, 1978.

Wood, Bryce. *The Making of the Good Neighbor Policy*. New York: Columbia University Press, 1961.

SELECTED ENGLISH-LANGUAGE PERIODICALS

Caracas *Daily Journal*. Started by Exxon Oil.

Latin American Regional Reports: Andean Group. Published in the United Kingdom. Focused coverage of Venezuela.

Latin American Times. Published in the United States. A new journal with ties to the democratic establishment in Latin America.

Latin American Weekly Report. Published by the same group as the *Regional Report*. Its perspective is more left-of-center.

SELECTED SPANISH-LANGUAGE BIBLIOGRAPHY ON VENEZUELA

Brewer-Carías, Allan R. *Régimen Jurídica de las Empresas Públicas en Venezuela*. Caracas: Centro Latinoamericano de Administración Para el Desarrollo, Editorial Arte, 1980.

Editorial Ateneo de Caracas. *Venezuela 1979: Examen Y Futuro*. Caracas: Ateneo de Caracas, 1980.

Essenfeld de Breuer, Senta. *La Cara Oculta del Desarrollo*. Caracas: Ediciones Conjuntas, Ateneo de Caracas, Jurídico de Venezuela, 1982.

Gil Yepes, José Antonio. *El Reto de las Elites*. Madrid: Tecnos, 1978.

Kaplan, Marcos, ed. *Petróleo y Desarrollo en México y Venezuela*. México: U.N.A.M., 1981.

Lander, Luís, and María Josefina de Rangel. *La Planificación en Venezuela*. Caracas: Sociedad Venezolana de Planificación, 1970.

Lestienne, Bernard. *El Sindicalismo Venezolano*. Caracas: Centro Gumilla, Curso de Formación Socio-Político #26, 1981.

Mayobre, José Antonio. *Las Inversiones Extranjeras en Venezuela*. Caracas: Monte Avila, 1970.

Mezger, Dorathea, ed. *Petróleo y Ecodesarrollo en Venezuela*. Caracas: Instituto de Investigaciones Sociales, Ateneo de Caracas, 1981.

Moreno León, José Ignacio. *Profundización de la Nacionalización Petrolera Venezolana: Aspectos Económicos, Administrativos y Fiscales*. Caracas: Ediciones Centauro, 1981.

Pérez Alfonzo, Juan Pablo, and Domingo Alberto Rangel. *El Desastre*. Valencia: Vadell Hermanos, 1976.

Pérez Alfonzo, Juan Pablo, and Ivan Loscher. *Alternativas*. Caracas: Garbizu & Todtmann, 1976.

Petkoff, Teodoro. *Democracia Para el Socialismo*. Caracas: Ediciones Sorocaima, 1981.

———. *¿Socialismo Para Venezuela?*. Caracas: Domingo Fuentes, 1970.

Rey, Juan Carlos, ed. *El Finaciamiento de los Partidos Políticos y la Democracia en Venezuela*. Caracas: Instituto de Estudios Políticos, Universidad Central de Venezuela, 1981.

Salinger, Pierre. *Cuadernos Venezolanos: Viajes y Conversaciones con Carlos Andrés Pérez*. Barcelona, Spain: Seix Barral, 1978.

Segura, Luis Buitrago. *Caracas, La Horrible*. Caracas: Ateneo de Caracas, 1980.

Tinoco, Pedro. *El Estado Eficaz*. Caracas: Colección los Desarrollistas #8, undated.

Travies, Fernando. *Ciudad, Región y Subdesarrollo*. Caracas: Fondo Editorial Comun, undated.

SELECTED VENEZUELAN PERIODICALS

El Diario de Caracas. A quality independent newspaper also available in the Library of Congress.

El Nacional. A quality Caracas newspaper available in the Library of Congress. It has an excellent op-ed page and a superb political cartoonist, "Zapata."

El Universal. A quality probusiness newspaper available in the Library of Congress.

Resúmen. A quality investigative weekly magazine. Its editor has been arrested twice by overly sensitive presidents. Scattered issues are available in the Library of Congress.

INDEX

NAME

SUBJECT